8/20/96

INTERNATIONAL UNION OF CRYSTALLOGRAPHY
MONOGRAPHS ON CRYSTALLOGRAPHY

D0139472

INTERNATIONAL UNION OF CRYSTALLOGRAPHY
BOOK SERIES

This volume forms part of a series of books sponsored by the International Union of Crystallography (IUCr) and published by Oxford University Press. There are three IUCr series: IUCr Monographs on Crystallography, which are in-depth expositions of specialized topics in crystallography; IUCr Texts on Crystallography, which are more general works intended to make crystallographic insights available to a wider audience than the community of crystallographers themselves; and IUCr Crystallographic Symposia, which are essentially the edited proceedings of workshops or similar meetings supported by the IUCr.

IUCr Monographs on Crystallography

1 *Accurate molecular structures: Their determination and importance*
   A. Domenicano and I. Hargittai, *editors*
2 *P. P. Ewald and his dynamical theory of X-ray diffraction*
   D. W. J. Cruickshank, H. J. Juretschke, and N. Kato, *editors*
3 *Electron diffraction techniques, Volume 1*
   J. M. Cowley, *editor*
4 *Electron diffraction techniques, Volume 2*
   J. M. Cowley, *editor*
5 *The Rietveld method*
   R. A. Young, *editor*

IUCr Texts on Crystallography

1 *The solid state: From superconductors to superalloys*
   A. Guinier and R. Jullien, *translated by* W. J. Duffin
2 *Fundamentals of crystallography*
   C. Giacovazzo, *editor*

IUCr Crystallographic Symposia

1 *Patterson and Pattersons: Fifty years of the Patterson function*
   J. P. Glusker, B. K. Patterson, and M. Rossi, *editors*
2 *Molecular structure: Chemical reactivity and biological activity*
   J. J. Stezowski, J. Huang, and M. Shao, *editors*
3 *Crystallographic computing 4: Techniques and new technologies*
   N. W. Isaacs and M. R. Taylor, *editors*
4 *Organic crystal chemistry*
   J. Garbarczyk and D. W. Jones, *editors*
5 *Crystallographic computing 5: From chemistry to biology*
   D. Moras, A. D. Podjarny, and J. C. Thierry, *editors*
6 *Crystallographic computing 6: A window on modern crystallography*
   H. D. Flack, L. Parkanyi, and K. Simon, *editors*
7 *Correlations, transformations, and interactions in organic crystal chemistry*
   D. W. Jones and A. Katrusiak, *editors*

The Rietveld Method

Edited by

R. A. YOUNG

Professor Emeritus, School of Physics
Georgia Institute of Technology, Atlanta

INTERNATIONAL UNION OF CRYSTALLOGRAPHY
OXFORD UNIVERSITY PRESS
1995

Oxford University Press, Walton Street, Oxford OX2 6DP

Oxford   New York
Athens   Auckland   Bangkok   Bombay
Calcutta   Cape Town   Dar es Salaam   Delhi
Florence   Hong Kong   Istanbul   Karachi
Kuala Lumpur   Madras   Madrid   Melbourne
Mexico City   Nairobi   Paris   Singapore
Taipei   Tokyo   Toronto
and associated companies in
Berlin   Ibadan

Oxford is a trade mark of Oxford University Press

Published in the United States
by Oxford University Press Inc., New York

© Oxford University Press, 1993
First published 1993
First published in paperback 1995

A catalogue record for this book is available from the British Library

Library of Congress Cataloging in Publication Data
The Rietveld method / edited by R. A. Young
(International Union of Crystallography monographs
on crystallography ; 5)
1. Rietveld method.   2. X-rays—Diffraction—Data processing.
I. Young, R. A. (Robert Alan), 1921–   .   II. Series.
QD945.R53   1993   548'.83—dc20   92-20900
ISBN 0 19 855912 7

Typeset by Integral Typesetting, Great Yarmouth, Norfolk
Printed in Great Britain by
Biddles Ltd, Guildford and King's Lynn

Preface

The Rietveld method is now widely recognized to be uniquely valuable for structural analyses of nearly all classes of crystalline materials not available as single crystals. This book is intended primarily to have tutorial and advisory value for those who already have some experience with this important method, but with Chapter 1 the beginner is also welcomed.

This book is distinctly not a 'proceedings' but it is a product of the same thrust that was responsible for the International Workshop on the Rietveld Method hosted by the Netherlands Energy Research Foundation ECN in Petten, The Netherlands, 13–15 June 1989. The Workshop was organized by the IUCr Commission on Powder Diffraction (CPD) in co-operation with the ECN and was sponsored by the International Union of Crystallography (IUCr). The CPD was established during the August 1987 IUCr meeting in Perth and, recognizing the need and the importance, undertook to organize the Workshop and to encourage the outgrowth of a book from it as one of its first, highest priority projects.

All invited lecturers at the Workshop were also invited to contribute a chapter for a book that would be much more coherent and internally consistent than a 'proceedings' would be. Not all chose to accept, perhaps in part because of the long delay that would be involved, first in getting the manuscript drafts from all authors and second in the revision procedure needed to achieve coherence and consistency.

This drive for coherence and consistency entailed considerable additional work on the parts of the chapter authors in revising their manuscripts to take account of what the other authors were saying and to unify, as much as feasible, the symbols and terms used for the same things. It also involved a considerable passage of time. A result is that the average date of currency of the material presented is probably late 1990 to early 1991, rather than June 1989.

In this drive for consistency, an effort has been made to reach a reconciliation of differing views so that the reader may be confident that what is said in this book is relatively authoritative and not likely to receive serious intellectual challenge. For the most part this has been achieved almost completely, thanks to the splendid co-operation and hard work of the authors. However, in rare cases, some differing strong opinions held by some authors could not be melded into the whole. When the editor could not reconcile the differences, the benefit of the doubt was given the author and his opinion is printed as presented. The reader is therefore advised that, in

spite of the editor's best efforts, in rare instances some highly individual
opinions may be expressed in some of the chapters. 'Recommendations', in
particular, are often subjective and not espoused by all authors. For the most
part, however, reconciliation of views and elimination of redundancy has
been accomplished and the editor is very grateful to the authors for their
patience and extra efforts which have led to this desired result.

December 1991   R. A. Young
Atlanta

Contents

Contributors

Baerlocher, Christian, Institut für Kristallographie und Petrographie, ETH Zentrum, CH-8092 Zürich, Switzerland

Cheetham, Anthony K., Materials Department, University of California, Santa Barbara, CA 93106, USA

David, W. I. F., Neutron Science Division, Rutherford Appleton Laboratory, Chilton, Didcot, Oxon OX11 0QX, England

de Keijser, Thomas H., Laboratory of Metallurgy, Delft University of Technology, Rotterdamseweg 137, 2628 AL Delft, The Netherlands

Delhez, Robert, Laboratory of Metallurgy, Delft University of Technology, Rotterdamseweg 137, 2628 AL Delft, The Netherlands

Hill, Roderick J., CSIRO Division of Mineral Products, PO Box 124, Port Melbourne, Victoria 3207, Australia

Izumi, F., National Institute for Research in Inorganic Materials, 1-1 Namiki, Tsukuba, Ibaraki 305, Japan

Jorgensen, James D., Materials Science Division, Argonne National Laboratory, Argonne, IL 60439, USA

Langford, J. Ian, School of Physics and Space Research, University of Birmingham, Birmingham B15 2TT, England

Louër, Daniel, Laboratoire de Cristallochimie, Université de Rennes I, Avenue du Général Leclerc, 35042 Rennes Cédex, France

Mittemeijer, Eric J., Laboratory of Metallurgy, Delft University of Technology, Rotterdamseweg 137, 2628 AL Delft, The Netherlands

Prince, Edward, Reactor Radiation Division, Materials Science and Engineering Laboratory, National Institute of Standards and Technology, Gaithersburg, MD 20899, USA

Richardson Jr, James W., Intense Pulsed Neutron Source Division, Argonne National Laboratory, Argonne, IL 60439, USA

Rietveld, Hugo M., Netherlands Energy Research Foundation ECN, PO Box 1, 1755 ZG Petten, The Netherlands

Sabine, Terence M., Advanced Materials Program, ANSTO, Menai, New South Wales 2234, Australia

Sonneveld, Eduard J., Laboratory of Metallurgy, Delft University of Technology, Rotterdamsweweg 137, 2628 AL Delft, The Netherlands

Snyder, Robert L., Institute for Ceramic Superconductivity, New York State College of Ceramics, Alfred University, Alfred, NY 14802-1296, USA

Suortti, P., ESRF, BP 220, F-38043 Grenoble Cédex, France

Toraya, Hideo, Ceramics Research Laboratory, Nagoya Institute of Technology, Asahigaoka, Tajimi 507, Japan

Von Dreele, Robert B., LANSCE, MS H805, Los Alamos National Laboratory, Los Alamos, NM 87545, USA

Young, R. A., School of Physics, Georgia Institute of Technology, Atlanta, GA 30332-0430, USA

1

Introduction to the Rietveld method

R. A. Young

1.1 Overview

1.1.1 *Naming and key features of the Rietveld method*

In Chapter 2, Dr Rietveld has provided an account of the inception and early development of his method and of the growth of its acceptance from being initially ignored to the very widespread and vigorous use in many fields that it now enjoys. In this chapter, we provide a general introductory account of what the method is and how it works.

First, why do we prefer to call this method the Rietveld method rather than 'profile refinement' or, better, whole-pattern-fitting structure refinement (PFSR)? The answer to the first part is that the principal goal is to refine crystal structures, not profiles. The things actually being refined are parameters in models for the structure and for other specimen and instrument effects on the diffraction pattern. The answer to the second part of the question has both a trivial part and an important part. The trivial part is that PFSR is a clumsy name. The important part of the answer is that it is primarily because of Rietveld's work and his open-handed sharing of all aspects of it that the last two decades have seen great advances in our ability to extract detailed crystal structural information from powder diffraction data, e.g. to do systematic structure refinement. In the mid-sixties, it became apparent to various diffractionists that much more information could be obtained from a powder pattern if the full power of computers could be applied to full-pattern analysis. The recognized point was that in a step-scanned pattern, for example, there was some information attached to each intensity at each step in the pattern, even if it were the negative information that there was no Bragg-reflection intensity there or the partial and initially scrambled information that the intensity at a step was the sum of contributions from the details of several Bragg reflections. It was Rietveld (i) who first worked out computer-based analytical procedures (quite sophisticated ones for the time, as it turned out) to make use of the full information content of the powder pattern, (ii) who put them in the public domain by publication of two seminal papers (1967, 1969) and (iii) who, very importantly, freely and widely shared his computer program. It is for these reasons that the method is appropriately referred to now as 'the Rietveld method', or 'Rietveld refinement', or 'Rietveld analysis'.

In the Rietveld method the least-squares refinements are carried out until the best fit is obtained between the entire observed powder diffraction pattern taken as a whole and the entire calculated pattern based on the simultaneously refined models for the crystal structure(s), diffraction optics effects, instrumental factors, and other specimen characteristics (e.g. lattice parameters) as may be desired and can be modelled. A key feature is the feedback, during refinement, between improving knowledge of the structure and improving allocation of observed intensity to partially overlapping individual Bragg reflections.

This key feature is not present in other popular methods of structure refinement from powder data in which (i) the assignment of all observed intensity to individual Bragg reflections and (ii) subsequent structure refinement with the so-derived Bragg intensities are carried out as separate, non-interacting procedures. The first procedure is appropriately called 'pattern decomposition' and may be defined as a systematic procedure for decomposing a powder pattern into its component Bragg reflections without reference to a crystal structural model. It is the subject of Chapter 14. The decomposition procedures in general use may be classified into two categories according to whether the lattice parameters are known and are used to fix the positions of the possible Bragg reflections. Such use of the lattice parameters goes a long way toward reducing the intensity-assignment ambiguity to acceptable levels. Among the particularly informative papers in the pattern-decomposition and non-Rietveld structure refinement areas are those by Will (1979), Pawley (1981), Will *et al.* (1983), Langford, *et al.* (1986), and Langford and Louër (1991), and, notably, Chapter 14, here, by Toraya.

1.1.2 *Core mathematics and procedures*

Obviously, the powder data must be in digitized form in order to be used in today's computers. Hence the diffraction pattern is recorded in digitized form, i.e. as a numerical intensity value, y_i, at each of several thousand equal

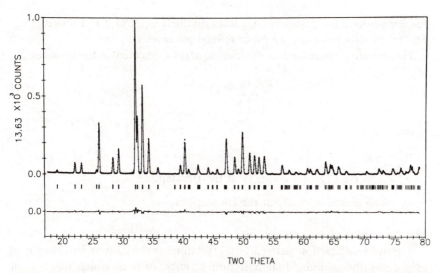

Fig. 1.1 Example of a Rietveld refinement plot. The specimen was fluorapatite. The observed intensity data, y_i, are plotted in the upper field as points with vertical error bars representing the counting statistical e.s.d.'s in them. The calculated pattern is shown in the same field as a solid-line curve. The difference, observed minus calculated, is shown in the lower field. The short vertical bars in the middle field indicate the positions of possible Bragg reflections.

increments (steps), i, in the pattern. Depending on the method, the increments may be either in scattering angle, 2θ, or some energy parameter such as velocity (for time-of-flight neutron data) or wavelength (for X-ray data collected with an energy dispersive detector and an incident beam of 'white' X-radiation). For constant wavelength data, the increments are usually steps in scattering angle and the intensity y_i at each step i, in the pattern is measured either directly with a quantum detector on a diffractometer or indirectly with step-scanning (digitizing) microdensitometry of film (e.g. Guinier) data. Typical step sizes range from 0.01 to 0.05° 2θ for fixed-wavelength X-ray data and a bit larger for fixed wavelength neutron data. Rietveld refinement with neutron diffraction data has been notably successful both with fixed-wavelength data, with which Rietveld (Chapter 2) developed and used the method, and with TOF (energy-dispersion via velocity discrimination) data (see Chapter 11). In any case, however, the number of steps in the powder diffraction pattern is usually in the thousands. Further, the Rietveld method, itself, is the same no matter what powder diffraction data are used. The differences among data sources affect the data preparation that is required, whether the steps are in angle or energy, and the instrumental parameters that are refined but not in the method itself.

In all cases, the 'best-fit' sought is the best least-squares fit to all of the thousands of y_i's simultaneously. Figure 1.1 shows an example of the plots resulting from a test case involving fluorapatite $(Ca_{10}(PO_4)_6F_2)$, space group $P6_3/m$, 19 structural and 15 other refined parameters.

The quantity minimized in the least-squares refinement is the residual, S_y:

$$S_y = \sum_i w_i(y_i - y_{ci})^2 \qquad (1.1)$$

where

   $w_i = 1/y_i$,

   $y_i =$ observed (gross) intensity at the ith step,

   $y_{ci} =$ calculated intensity at the ith step,

and the sum is overall data points.

A powder diffraction pattern of a crystalline material may be thought of as a collection of individual reflection profiles, each of which has a peak height, a peak position, a breadth, tails which decay gradually with distance from the peak position, and an integrated area which is proportional to the Bragg intensity, I_K, where K stands for the Miller indices, h, k, l. I_K is proportional to the square of the absolute value of the structure factor, $|F_K|^2$. In all powder diffraction patterns but those so simple that the Rietveld method is not needed in the first place, these profiles are not all resolved but partially overlap one another to a substantial degree.

It is a crucial feature of the Rietveld method that no effort is made in advance to allocate observed intensity to particular Bragg reflections nor to resolve overlapped reflections. Consequently, a reasonably good starting model is needed. The method is a structure refinement method. It is not a structure solution method, *per se*, though it can be a very important part of a structure solution package. (See Chapter 15.)

Typically, many Bragg reflections contribute to the intensity, y_i, observed at any arbitrarily chosen point, i, in the pattern. The calculated intensities y_{ci} are determined from the $|F_K|^2$ values calculated from the structural model by summing of the calculated contributions from neighbouring (i.e. within a specified range) Bragg reflections plus the background:

$$y_{ci} = s \sum_K L_K |F_K|^2 \phi(2\theta_i - 2\theta_K)P_K A + y_{bi} \qquad (1.2)$$

where

   s is the scale factor,

   K represents the Miller indices, h k l, for a Bragg reflection,

   L_K contains the Lorentz, polarization, and multiplicity factors,

   ϕ is the reflection profile function (see Section 1.3 and Table 1.2),

P_K is the preferred orientation function (see Section 1.3),
A is an absorption factor,
F_K is the structure factor for the Kth Bragg reflection (see Section 1.3), and
y_{bi} is the background intensity at the ith step.

The effective absorption factor, A, differs with instrument geometry. It is elucidated in Chapter 4. It is usually taken to be a constant for the instrument geometry most used for X-ray diffractometers, that of a flat specimen with its surface maintained normal to the diffraction vector by a θ–2θ relationship between specimen rotation and detector rotation about the diffractometer axis. It does depend on angle for other geometries.

In a number of available computer programs for the Rietveld method, the ratio of the intensities for the two X-ray K_α wavelengths (if used) is absorbed in the calculation of $|F_K|^2$, so that only a single scale factor is required.

The least squares minimization procedures lead to a set of *normal equations* involving derivatives of all of the calculated intensities, y_{ci}, with respect to each adjustable parameter and are soluble by inversion of the *normal matrix* with elements M_{jk} formally given by

$$M_{jk} = -\sum_i 2w_i\left[(y_i - y_{ci})\frac{\partial^2 y_{ci}}{\partial x_j\,\partial x_k} - \left(\frac{\partial y_{ci}}{\partial x_j}\right)\left(\frac{\partial y_{ci}}{\partial x_k}\right)\right] \qquad (1.3)$$

where the parameters x_j, x_k are the (same set of) adjustable parameters. In the use of this algorithm, it is common practice to approximate these matrix elements by deletion of the first term, that in $(y_i - y_{ci})$.

One is, thus, dealing with the creation and inversion of an m by m matrix, where m is the number of parameters being refined. Because the residual function is non-linear, the solution must be found with an iterative procedure in which the shifts, Δx_k, are

$$\Delta x_k = \sum M_{jk}^{-1}\frac{\partial S_y}{\partial x_k}. \qquad (1.4)$$

The calculated shifts are applied to the initial parameters to produce a, supposedly, improved model and the whole procedure is then repeated. Because the relationships between the adjustable parameters and the intensities are non-linear, the starting model must be close to the correct model or the non-linear least squares procedure will not lead to the global minimum. Rather, the procedure will either diverge or lead to a false minimum if the starting point is in its domain. (This is true for all non-linear least-squares refinements, not just Rietveld refinements.) Selection of different least-squares algorithms at different stages of refinement may alleviate the false-minimum problem in some cases (see Chapter 13). Another approach is to use multiple data sets of different kinds, e.g. X-ray and neutron, simultaneously (Chapter 12) or to use constraints (Chapter 10). Other approaches can also help one to avoid false minima or, at least, to be properly suspicious that a minimum may not necessarily be the global minimum just because it is stable. Chapter 3 contains the definitive and more extensive discussion of these mathematical matters.

The model parameters that may be refined include not only atom positional, thermal, and site-occupancy parameters but also parameters for the background, lattice, instrumental geometrical–optical features, specimen aberrations (e.g. specimen displacement and transparency), an amorphous component (Chapter 6), and specimen reflection-profile-broadening agents such as crystallite size and microstrain (Chapters 7, 8, 9, and 11). In some cases it is important to model extinction, as well. Although it is in general a much less severe problem with powders than with single crystals, extinction can be quite important in some powder specimens (Chapter 4). Multiple phases may be refined simultaneously and comparative analysis of the separate overall scale factors for the phases offers what is probably the most reliable current method for doing quantitative phase analysis. (See Appendix 5.A to Chapter 5.) The usual refinable parameters are listed by class in Table 1.1. Individual anisotropic thermal parameters have rarely been refined satisfactorily from ordinary laboratory X-ray powder diffraction data, but they have been with fixed wavelength neutron data and are refined almost routinely with TOF neutron powder diffraction data.

1.1.3   Examination of the factors y_{bi}, ϕ, P_K, and F_K in eq. (1.2)

The *background intensity* at the ith step, y_{bi}, may be obtained from (i) an operator-supplied table of background intensities, or (ii) linear interpolation between operator-selected points in the pattern, or (iii) a specified background function. There is a growing general belief in the field that the background should always be refined unless there is a specific reason to the contrary, e.g. pattern simplicity. If the background is to be refined, y_{bi} must be obtained from a refinable background function which may be phenomenological or, better, based on physical reality and include refinable models for such things as an amorphous component (e.g. via a radial distribution function, see Chapter 6) and thermal diffuse scattering (TDS) contributions as calculated in Chapter 9. A simple phenomenological function which has been useful in the absence of a better one is a fifth-order polynomial provided with an operator-specifiable origin to allow added flexibility in fitting broad humps in the background curve,

$$y_{bi} = \sum_{m=0}^{5} B_m [(2\theta_i / \text{BKPOS}) - 1]^m, \tag{1.5}$$

Table 1.1   Parameters refinable simultaneously

For *each* phase present (e.g. 1–8):

x_j y_j z_j B_j N_j

(x_j, y_j, and z_j are position coordinates, B_j is an isotropic thermal parameter, and N_j is the site-occupancy multiplier, all for the jth atom in the unit cell.)

Scale factor (note quantitative phase analysis possibility)
Specimen-profile breadth parameters (e.g. 1–5)
Lattice parameters
Overall temperatue factor (thermal parameter)
Individual anisotropic thermal parameters
Preferred orientation
Crystallite size and microstrain (through profile parameters)
Extinction

Global

2θ-Zero
Instrumental profile
Profile asymmetry
Background
    (Parameters in analytical function, e.g. 5th order polynomial, sum of exponentials,
    RDF for amorphous component, etc.)
Wavelength
Specimen displacement
Specimen transparency
Absorption

where BKPOS is the origin that is to be user-specified in the input control file.

The *reflection profile function*, ϕ, approximates the effects of both instrumental features (including reflection-profile asymmetry) and, possibly, specimen features such as aberrations due to absorption (transparency), specimen displacement, and specimen-caused broadening (e.g. by crystallite-size and microstrain effects) of the reflection profiles (see Chapters 7, 8, 9, and 11). For energy dispersive data, e.g. collected with neutron time-of-flight (TOF) or synchrotron 'white' radiation and energy sensitive detector techniques, 2θ, must be replaced by an energy or monentum variable, e.g. $(4\pi \sin \theta)/\lambda$ (= 'k' or 'q'), or time-of-flight, etc. Analytical reflection profile functions available in some of the most widely used programs include two different pseudo-Voigt functions, the Pearson VII function, and Gaussian,

Lorentzian, and modified Lorentzian functions. They are listed in Table 1.2. Other programs offer refinements on these plus additional, or different, choices including differentiable numeric representations (see Baerlocher's program listed in Section 1.6). Perhaps of particular interest are the 'split-Pearson VII' functions used advantageously to model reflection profile asymmetry (see Chapter 7) and the Voigt function described in Chapter 8.

For angle dispersive data, the dependence of the breadth H (also called Γ in this book) of the reflection profiles measured as full-width-at-half-maximum (FWHM) has typically been modelled as (Caglioti *et al.* (1958)

$$H^2 = U \tan^2 \theta + V \tan \theta + W \tag{1.6}$$

where U, V, and W are the refinable parameters. Initially developed for 'medium' (or less) resolution neutron powder diffractometers, it worked quite satisfactorily for them, as did simple Gaussian reflection profile functions. Even though the instrumental diffraction profiles of the typical X-ray powder diffractometers operating on sealed-off X-ray tube or rotating anode sources are generally neither Gaussian nor symmetric, the Caglioti *et al.* relation has been widely used for lack of anything better that was sufficiently simple. With the reflection profiles from X-ray diffractometers and other comparatively high resolution instruments, such as Guinier cameras and genuinely high-resolution neutron powder diffractometers, another complication arises. Their instrumental profiles are sufficiently narrow so that broadening of the intrinsic diffraction profile from specimen defects such as microstrain and small crystallite size is now a significant part of the total and it does not have the Caglioti *et al.* dependence on angle. The separate broadening contributions do not simply add, but the way they do combine has been well known for more than fifty years. (See Fig. 1 of Jones 1938, and eqn 1 of Stokes 1948 plus his ensuing discussion of Jones's work. The author is grateful to Professor A. J. C. Wilson for discussions of this point.) It is by the process known mathematically as convolution:

$$h(x) = \int g(x')f(x - x')\,\mathrm{d}x' \tag{1.7}$$

where x is $(2\theta_i - 2\theta_K)$ (as in eqn 1.2), x' is the variable of integration in the same x domain, $g(x)$ is the instrumental profile function, $f(x)$ is the intrinsic diffraction ('specimen broadened') profile function, and $h(x)$ is the resulting observed profile function. Clearly, as long as $g(x)$ is much broader than $f(x)$, the observed profile function will reflect little or no more of the character of $f(x)$ than it would if $f(x)$ were a true delta function.

If the intrinsic diffraction profile, $f(x)$, is not narrow compared to $g(x)$, then it must be modelled at least approximately so that the profile fits will

Table 1.2 Some symmetric analytical profile functions that have been used[a]

Function	Name
(a) $\dfrac{C_0^{1/2}}{H_K \pi^{1/2}} \exp(-C_0(2\theta_i - 2\theta_K)^2/H_K^2)$	Gaussian ('G')
(b) $\dfrac{C_1^{1/2}}{\pi H_K} 1 \Big/ \left[1 + C_1 \dfrac{(2\theta_i - 2\theta_K)^2}{H_K^2} \right]$	Lorentzian ('L')
(c) $\dfrac{2C_2^{1/2}}{\pi H_K} 1 \Big/ \left[1 + C_2 \dfrac{(2\theta_i - 2\theta_K)^2}{H_K^2} \right]^2$	Mod 1 Lorentzian
(d) $\dfrac{C_3^{1/2}}{2H_K} 1 \Big/ \left[1 + C_3 \dfrac{(2\theta_i - 2\theta_K)^2}{H_K^2} \right]^{3/2}$	Mod 2 Lorentzian
(e) $\eta L + (1 - \eta)G$	pseudo-Voigt ('pV')

The mixing parameter, η, can be refined as a linear function of 2θ wherein the refinable variables are NA and NB:

$$\eta = NA + NB*(2\theta)$$

Function	Name
(f) $\dfrac{C_4}{H_K} \left[1 + 4*(2^{1/m} - 1) \dfrac{(2\theta_i - 2\theta_K)^2}{H_K^2} \right]^{-m}$	Pearson VII

m can be refined as a function of 2θ,

$$m = NA + NB/2\theta + NC/(2\theta)^2,$$

where the refinable variables are NA, NB, and NC.

(g) Modified Thompson–Cox–Hastings pseudo-Voigt,   (Mod-TCH pV)
   'TCHZ'

$$\text{TCHZ} = \eta L + (1 - \eta)G$$

where

$$\eta = 1.36603q - 0.47719q^2 + 0.1116q^3$$

$$q = \Gamma_L/\Gamma$$

$$\Gamma = (\Gamma_G^5 + A\Gamma_G^4\Gamma_L + B\Gamma_G^3\Gamma_L^2 + C\Gamma_G^2\Gamma_L^3 + D\Gamma_G\Gamma_L^4 + \Gamma_L^5)^{0.2} = H_K$$

$$\begin{aligned} A &= 2.69269 & B &= 2.42843 \\ C &= 4.47163 & D &= 0.07842 \end{aligned}$$

$$\Gamma_G = (U \tan^2\theta + V \tan\theta + W + Z/\cos^2\theta)^{1/2}$$

$$\Gamma_L = X \tan\theta + Y/\cos\theta$$

[The modification consists of adding the term with the parameter Z to the original Thompson et al. (1987) pseudo-Voigt function ('TCH') in order to provide a component of the Gaussian FWHM which is constant in $\mathbf{d^*}$, as is the Y component of the Lorentzian FWHM.]

Table 1.2   (*continued*)

In the above profile functions,

H_K is the full-width-at-half-maximum (FWHM) of the Kth Bragg reflection, as is Γ,

the refinable parameters are those in η, m, and H or Γ,

$C_0 = 4 \ln 2$,

$C_1 = 4$,

$C_2 = 4(2^{1/2} - 1)$,

$C_3 = 4(2^{2/3} - 1)$, and

$$C_4 = \frac{2\sqrt{m}\,(2^{1/m} - 1)^{1/2}}{(m - 0.5)\,\pi^{1/2}}.$$

[a] The origins and performances of most of these functions are discussed in Young and Wiles (1982). The TCHZ function is discussed in Young and Desai (1989).

be good enough to support valid estimates of the Bragg intensities, even in Rietveld refinement. Modelling them better than approximately can support determination of microstructural information, principally crystallite size and microstrain parameters. Asymmetry and anisotropic (i.e. dependent on crystallographic direction as well as scattering angle) effects on the $f(x)$ profile continue to be particular problems in the fitting of calculated to observed reflection profiles, but progress is being made. Chapters 7, 8, 9, and 11 contain much more information on these topics. An excellent example of the determination of anisotropic crystallite microstrain is given in Chapter 11, done in that case with TOF neutron powder diffraction data.

Preferred orientation arises when there is a stronger tendency for the crystallites in a specimen to be oriented more one way, or one set of ways, than all others. An easily visualized case of preferred orientation is that which results when a material with a strong cleavage or growth habit is packed into a flat specimen holder backed by a glass slide. The cleavage (or growth) faces, all of the same crystallographic type, tend to be aligned parallel to the flat backing surface. Ordinary processed table salt provides an example. It tends to re-crystallize, during processing, in little parallelepipeds (often nearly cubic) bounded by {001} faces. When it is packed into a flat holder for a standard θ–2θ diffractometer, the {001} faces will tend to be oriented parallel to the specimen surface more than any other way and the result will be that the {001} diffracted intensities will be disproportionately strong. Removing and repacking the specimen with the same procedure may change the degree of disproportionate strength of the {001} reflections but will not change the fact of it. Preferred orientation is normally less of a problem in neutron than in X-ray powder diffraction because of the larger specimens;

a larger fraction of the diffracting crystallites are farther from the bounding surfaces and their orientations are therefore less influenced by those surfaces.

Preferred orientation should not be confused with 'graininess' (also referred to as an inadequate 'powder average'), the situation in which there are so few crystallites being irradiated that the number of crystallites correctly oriented to diffract differs significantly in a random fashion from reflection to reflection of different types. In this latter case, there is no lack of randomness in the choice of orientation of any given crystallite and unloading and repacking the specimen will lead to a different set of relative intensities. Readers familiar with a reciprocal space view of powder diffraction, will recognize that this graininess is easily visualized as a non-smoothly varying density (i.e. grainy character) in the shells formed, one for each hkl, in the reciprocal space of a powder. There is no way to make an after-the-fact correction for this problem.

Because preferred orientation produces systematic distortions of the reflection intensities, after-the-fact corrections can be made for it, i.e. the distortions can be mathematically modelled with 'preferred orientation functions', P_K in eqn (1.2). P_K was implemented in Rietveld's own early programs and, until rather recently, in most others since as

$$P_K = \exp(-G_1 \alpha_K^2) \tag{1.8}$$

or

$$P_K = (G_2 + (1 - G_2) \exp(-G_1 \alpha_K^2) \tag{1.9}$$

where G_1 and G_2 are refinable parameters and α_K is the angle between $\mathbf{d}_K^*$ and the fibre axis direction. If not already present, the assumed axial symmetry about the diffraction vector can be provided by spinning the sample. This P_K is useful if the degree of preferred orientation is not large. When this function is used in Rietveld refinements with data from a standard θ–2θ X-ray diffractometer, the refined parameter changes sign with a change from platey to acicular habit, but it still improves the fit.

Recently, Dollase (1986) showed the superior performance of the March function. Many authors have now incorporated the March–Dollase function in their Rietveld refinement codes and confirm Dollase's evaluation. Some of the particular merits of this preferred orientation function,

$$P_K = (G_1^2 \cos^2 \alpha + (1/G_1) \sin^2 \alpha)^{-3/2}, \tag{1.10}$$

are discussed in the appendix (5.A) to Chapter 5.

Another interesting and apparently very powerful preferred-orientation function is that of Ahtee et al. (1989). The preferred orientation effect is modelled by expanding the orientation distribution in spherical harmonics. They have incorporated that preferred orientation model in their new

Rietveld refinement code in which they have also used a true Voigt function for the reflection profile function. They state that 'the code is applicable to neutron, synchrotron and conventional X-ray data when the incident radiation is essentially one wavelength'. In tests with samples with textures known from pole figure measurements, they found that the corrections obtained from the refinement agreed 'very closely' with the measured values, even though the largest corrections exceeded a factor of two and only a few terms of the harmonic expansion needed to be used. It would seem that those who find cases in which the March–Dollase approach does not give satisfactory results might consider using this more complex but, in principle, more powerful approach.

The *structure factor*, F_K, is given by

$$F_K = \sum_j N_j f_j \exp[2\pi i(hx_j + ky_j + lz_j)] \exp[-M_j], \qquad (1.11)$$

where

h, k, and l are the Miller indices,

x_j, y_j, and z_j are the position parameters of the jth atom in the unit cell,

$$M_j = 8\pi^2 \overline{u_s^2} \sin^2 \theta/\lambda^2, \qquad (1.12)$$

$\overline{u_s^2}$ is the root-mean-square thermal (and random static) displacement of the jth atom parallel to the diffraction vector, and

$N_j =$ site occupancy multiplier for the jth atom site. N_j is the actual site occupancy divided by the site multiplicity. Note that if the scale factors of 2 or more phases are to be used for quantitative phase analysis (see Chapter 5, Appendix A), the site multiplicities must be correctly specified on the same basis in each phase, preferably an absolute basis. It is not sufficient for them to be correct on a separate relative basis within each phase.

Examination of eqns (1.2) and (1.11) can help one to appreciate the complementary relationship between Rietveld structure refinements done with X-ray data and those done with neutron data. The only really significant difference arises in the f_j factor in eqn (1.11) which, then, is inserted into eqn. (1.2). For this discussion, we set aside magnetic scattering of neutrons and deal only with nuclear scattering, which takes place by short-range nuclear forces. X-rays are scattered, almost only, by the electrons surrounding the nuclei. Thus, the region from which they are scattered, which is about 1–2 Å in diameter, is of the order of 10^4 larger than the region from which the neutrons are scattered. In each case, the scattering factor has the form of the Fourier transform (Guinier 1963) of the distribution of scattering density (density of the scattering material times its scattering power). The result is that the X-ray scattering factor falls off seriously over the range of $(\sin \theta)/\lambda$ in which we can make observations with the commonly used wavelengths in the range 0.5–2.5 Å. This is represented in Fig. 1.2. Note that the value of f_j at $\sin \theta/\lambda = 0$ is just the number of electrons, e.g. the atomic number, Z, for neutral atoms. (The units are those of scattering from a free 'Thomson' electron.) The Fourier transform of the much more compact region of scattering of the neutrons, however, does not fall off significantly at all over this normal range of observation. Thus, the atomic, really nuclear, scattering factors for (thermal) neutrons are effectively constants. The units are those of the square root of cross-section, i.e. 10^{-12} cm. It is customary to use b_j rather than f_j for this quantity, the *scattering length*. The b_j values differ in other important ways from the X-ray f_j's. Figure 1.3 (kindly provided by Professor H. Feuss, Frankfurt), exhibits many of these differences. (1) There is no regular progression with Z; (2) the b values can be *negative* (which means a 180 degree phase shift on scattering); (3) all b values (expressed in units of 10^{-12} cm) lie generally between $+1.5$ and -1.0 (^{164}Dy is the exception); (4) isotopes of the same element can, and generally do, have quite different b values. An interesting example of great practical utility is that of ^{1}H and ^{2}H ($=$D), for which the b's are -0.38 and $+0.65$, respectively, $\times 10^{-12}$ cm. Much use is made of these differing features in the practical application of the Rietveld method with neutron diffraction data.

1.2 Structural complexity treatable in Rietveld refinement

Fitch *et al.* (1982) reported successful refinement of 193 parameters, 168 atomic parameters for the main specimen, $UO_2DAsO_4 \cdot 4D_2O$, 12 for the ice that formed because of the low temperature (4 K), and 13 others (lattice, asymmetry, preferred orientation, overall scale, profile breadth). However, they refined the parameters in subsets so that the greatest number refined simultaneously was 161—which is still an impressively large number. The powder diffraction data were neutron data collected with instrument D1A at the ILL (Institut Max von Laue–Paul Langevin) in Grenoble, France. The resulting Rietveld refinement plot is shown in Fig. 1.4. There are 2098 Bragg reflections in this pattern. Because of overlap, this does not give quite the same significance to the over-determination ratio (>10) that one would have with 2098 single crystal reflections, but the effective over-determination is clearly adequate. Both the use of neutrons (rather than X-rays) and the low temperature contribute toward maintaining good intensities in the high angle region of the pattern, which is surely an advantage for refinement of such a complex structure.

Baerlocher (1984), none the less, has refined a 181 parameter problem with X-ray data on ZSM-5 zeolite. That is, the normal matrix was 181 by 181.

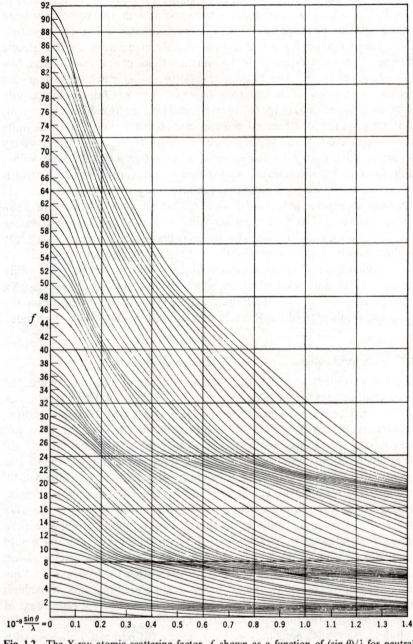

Fig. 1.2 The X-ray atomic scattering factor, f, shown as a function of $(\sin\theta)/\lambda$ for neutral atoms. The numbers along the ordinate are the atomic numbers, Z. (From Buerger 1942.)

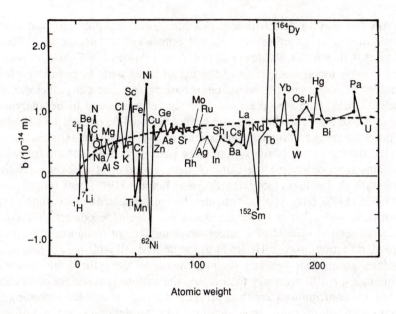

Fig. 1.3 Nuclear scattering lengths for thermal neutrons shown as a function of atomic weight. (Courtesy of Prof. H. Fuess.)

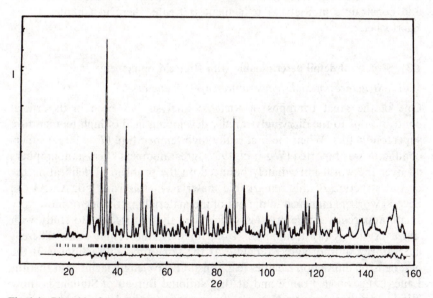

Fig. 1.4 Rietveld refinement plot from the 193 parameter problem (Section 1.2). Neutron data, 1.9 Å, $R_{wp} = 0.042$, $\Delta \approx 0.05$ Å and $\sigma \approx \frac{1}{2}\Delta$. (From Fitch *et al.* 1982.)

Another way to deal with complex structures, e.g. polymers and other macromolecular structures, in Rietveld refinement is with the use of constraints (Chapter 10). Pawley *et al.* (1977) and Pawley (1977) have provided particularly useful discussions of the use of constraints in powder pattern analysis. Various current computer programs for Rietveld refinement do include them. One idea is to constrain some parameters to be analytical functions of others so that only the second set need be refined. This is often a practical thing to do when one is refining a macromolecular structure in which the structures of the subunits are reasonably well known and what remains is to determine the distance between subunits and their relative orientations. Immirzi (1980), Immirzi and Iannelli (1987), and Iannelli and Immirzi (1989) have applied this idea in the refinement of isotactic polypropylene and other polymer structures for which the powder diffraction patterns tend to consist of a rather small number of recognizable, usually broad, reflection profiles. He refers to the desired refinable parameters, such as inter-group distances and torsion angles, as 'generalized coordinates'. Although it is still necessary to compute the scattering contribution of each atom, the contributions are then programmatically assembled according to their contributions to the generalized coordinates and the resulting normal-matrix size is only that of the number of generalized coordinates. The use of such a scheme for cellotetraose, for example, reduced the matrix size from > 100 to about 20 or less (depending on what things were assumed to be well enough known (Sakthivel *et al.* 1987). The use of restraints and constraints in Rietveld refinement is treated here in Chapter 10 by Baerlocher.

1.3 Structural detail determinable with Rietveld refinement

1.3.1 *Structures of high-temperature superconductors*

One of the great triumphs of Rietveld analysis has been in its crucial contributions to the dizzyingly rapidly developing field of high-temperature superconductors. When the first really high temperature ($T_c = 90$ K) superconductor was reported (Wu *et al.* 1987), great numbers of crystallographers all over the world immediately pounced on the problem of delineating the crystal structure of this remarkable material, in this case $YBa_2Cu_3O_{7-x}$. Several workers managed to dig out of the material matrix one or more very small, apparently single crystals which they then proceeded to study with single crystal X-ray techniques. Unfortunately, the results from different laboratories did not agree in detail. At the same time, diffractionists at the best neutron diffraction facilities (e.g. constant wavelength facilities at Institut Laue–Langevin in France and at the National Bureau of Standards (now NIST) and Brookhaven National Laboratory in the US; pulsed neutron (TOF) sources at the Rutherford–Appleton Laboratory in the UK, the IPNS at Argonne National Laboratory in the USA, and the pulsed neutron source at the National Laboratory for High Energy Physics at Tsukuba in Japan) studied larger samples of the polycrystalline matrix, performed Rietveld analysis with several different starting models, and all came to the same conclusion in detail. The structure was, thus correctly determined from powder diffraction data whereas the X-ray single crystal results had been in error. For partisans of the powder method, it is a rare chance to exult that 'powder beat single crystals'. The reason, however, is only because the 'single' crystals were not, in fact, single but were the victims of micro-twinning. Figure 12.5 (in Chapter 12) is a computer drawing of this structure taken from recent work of von Dreele and Larson (1988) resulting in 'the most accurate structure determination to date' for this remarkable material.

Neutron powder diffraction has since been the major player in crystal structural studies of the various other 'high T_c' materials since discovered and has gone far in elucidating the structural origins, or requirements, of superconductivity. One important parameter is the distribution of oxygen atoms between two incompletely filled sites. Figure 1.5 is a plot from the work of Jorgensen *et al.* (1987), which illustrates the detail determinable about the oxygen site occupancies. This plot is further discussed in Section 1.5. Another paper remarkable for the wealth of detail determined about a high T_c material is that of Williams *et al.* (1988), who used neutron and X-ray data simultaneously in the Rietveld refinement. (See also Chapter 12.) They obtained precise atom positions and anisotropic thermal parameters plus microstrain, and 'conclusively' demonstrated the absence of any cation disorder.

An example of another kind that shows the remarkable power of the Rietveld method to sort out details when the model truly fits the physical facts is given by Young and Sakthivel (1988). The 'data' were calculated according to models with random counting fluctuations added which were generated from a Gaussian distribution with a mean of zero and a standard deviation of 1. The 'experimental data set' was the sum of data so calculated from two models of $Nd_2FE_{14}B$ which differed only in the breadths of the pseudo-Voigt profiles used. One was three times broader than the other. Rietveld refinements were then carried out (starting with parameters including structural parameters displaced from those used in the calculation of the 'data'), first with a single-phase model and then with a two-phase model in which the two phases differed only in the starting profile breadths. One might speculate that this perfect overlap of every reflection with another differing only in profile breadth would cause the refinements to fail. The refinements with a single phase did, in fact, fail in the sense that they could not reproduce the parameters from which the data were calculated and did lead to a 'super Lorentzian' effect (Wertheim *et al.* 1974) in that the mixing parameter indicated more than 100 per cent Lorentzian character in the refined profile.

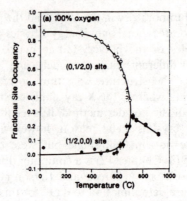

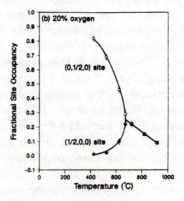

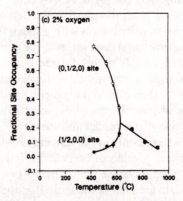

Fig. 1.5 Fractional 'O(1)' site occupancies at the indicated sites in $YBa_2Cu_3O_{7-x}$ as a function of temperature for a sample heated in atmospheres of the indicated oxygen contents. For the purpose of assessing precision, note how well the points fall on smooth curves. (From Jorgensen *et al.* 1987.)

When the two-phase model was used, however, the two sets of profile parameters used in the original calculation were faithfully recovered, as well as the crystal structural parameters, in spite of the perfect overlap and lack of any crystal-structural differences (except specimen broadening of the profiles) on which the separation could be based. This example makes it very clear that the Rietveld refinement method, *per se*, is capable of revealing far more detail than we customarily get from it; the limitations we currently face are in the models used and not in the method.

Additional recent, excellent examples of structural detail determined in Rietveld refinement are given in several of the chapters in this book. Perhaps particularly impressive are the examples in Chapters 11 and 12.

1.3.2 *Structural details determined in 'real time'*

A developing new dimension in Rietveld refinement is real-time studies of changing samples. This involves large numbers of data sets taken over a range of the sample variable(s), such as temperature, and automatic chaining of Rietveld refinements with the successive data sets. In recent years, there has been a strong thrust in such work at the Institut Laue–Langevin in France, a notably well-equipped and staffed laboratory with the best thermal neutron flux for diffraction studies in the world and a very large user base. An example is the work of Rodriguez-Carvajal *et al.* (1988) from which Fig. 1.6 is taken. The object was to reveal the structural reasons for an 'anomalous' change in strain in La_2NiO_4 accompanying an 80 K reversible phase transition in which the space group did not change. The question was whether it might be related to high temperature superconductivity, as magnetic ordering had been observed in nearly stoichiometric samples of this layered perovskite. The experiment was set up to collect data on structural changes in a 'real time' mode by collecting a full-powder diffraction pattern in every 3 min interval as the specimen temperature was increased linearly at 0.33 K min^{-1} from 3 to 275 K. A 400-cell position-sensitive detector was used on instrument D1B (the high-flux one) to collect 321 neutron powder diffraction patterns during this heating process. They used an on-site much-modified version of a DBWS Rietveld refinement code adapted to do sequential refinements with the 321 data sets automatically. The explanation for the occurrence of the strain without space group change was found in the changing tilt angles of the NiO_4 octahedra, as is detailed in Fig. 1.6 (their Fig. 4.). It is impressive to note that each point at a different location along the abscissa came from a different, complete Rietveld refinement made with an independent set of data. This example may be giving us a view into future problems with handling, storing, and doing refinements with great masses of different powder diffraction data sets. It is also giving us a view into a new world of relevant structural detail followed

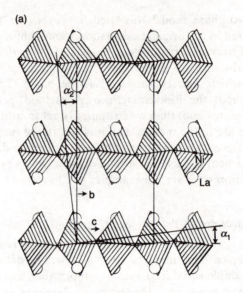

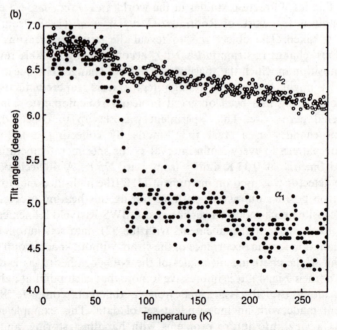

Fig. 1.6 (a) View of the structure of La_2NiO_4 along [100], showing the tilt angles of the NiO_4 octahedra. (b) Temperature dependence of the tilt angles α_1 and α_2. (Reprinted with permission from Rodriguez-Carvajal et al. 1988.)

on a sufficiently small time or temperature (or both) grid to give us a totally new quality of insight into the dynamic changes in structural details which, though small, have large effects on important material properties.

1.4 Criteria of fit

As has been noted, the Rietveld refinement process will adjust the refinable parameters until the residual (eqn 1.1) is minimized in some sense. That is, a 'best fit' of the entire calculated pattern to the entire observed pattern will be obtained. But the particular 'best fit' obtained will depend on the adequacy of the model (i.e. does it contain the parameters needed to model the actual structure and diffraction conditions 'well enough'?) and on whether a global minimum, rather than a local ('false') minimum, has been reached. One needs various criteria of fit in order to make these judgments. It is also important to have some kinds of indicators reported at each cycle so that one can judge whether the refinement is proceeding satisfactorily and when the refinement has become sufficiently near to completion that it can be stopped. A particularly authoritative discussion of most of these matters is given in Chapter 3. A lighter, more introductory presentation follows here.

Borrowing a page from the single-crystal crystallographers and adapting it to needs, the users of the Rietveld method have developed several R-values that are now commonly used (Table 1.3).

Because 'R-Bragg' and 'R-structure factor' are based not on actually observed Bragg intensities but on those deduced with the help of the model, they are, therefore, biased in favour of the model being used. None the less, they are the R's that are most nearly comparable to the conventional R-values quoted in the literature on single-crystal structure refinements. They also serve a useful function because they are insensitive to misfits in the pattern that do not involve the Bragg intensities of the phase(s) being modelled.

From a purely mathematical point of view, R_{wp} is the most meaningful of these R's because the numerator is the residual being minimized. For the same reason, it is also the one that best reflects the progress of the refinement.

Another useful numerical criterion is the 'goodness of fit', S (not to be confused with the residual, S_y), shown in Table 1.3. An S value of 1.3 or less is usually considered to be quite satisfactory. An S value of 1.7, for example, probably is a warning that you should look further into the reasons and question the adequacy of your model. On the other hand, a small S may simply mean that the counting statistical errors far outweigh the model errors either because of poor counting statistics or because of high background which, being slowly varying in angle, is easily modelled. In the first case, a data set with better counting statistics would permit improved refinement of model details to better approximate the physical facts. As Prince (author

Table 1.3 Some often-used numerical criteria of fit

$$R_F = \frac{\sum |(I_K(\text{`obs'}))^{1/2} - (I_K(\text{calc})^{1/2})|}{\sum (I_K(\text{`obs'}))^{1/2}}$$ ('R-structure factor')

$$R_B = \frac{\sum |I_K(\text{`obs'}) - I_K(\text{calc})|}{\sum I_K(\text{`obs'})}$$ ('R-Bragg factor')

$$R_p = \frac{\sum |y_i(\text{obs}) - y_i(\text{calc})|}{\sum y_i(\text{obs})}$$ ('R-pattern')

$$R_{wp} = \left\{ \frac{\sum w_i(y_i(\text{obs}) - y_i(\text{calc}))^2}{\sum w_i(y_i(\text{obs}))^2} \right\}^{1/2}$$ ('R-weighted pattern')

Here I_K is the intensity assigned to the Kth Bragg reflection at the end of the refinement cycles. In the expressions for R_F and R_B the 'obs' (for observed) is put in quotation marks because the Bragg intensity, I_K, is rarely observed directly; instead the I_K values are obtained from programmatic allocation of the total observed intensity in a 'scramble' of overlapped reflections to the individual reflections, according to the ratios of those reflection intensities in the calculated pattern.

The '*Goodness-of-fit*' indicator, S, is

$$S = [S_y/(N - P)]^{1/2} = R_{wp}/R_e$$

where

$$R_e = \text{`}R\text{-expected'} = [(N - P)/\sum w_i y_{oi}^2]^{1/2}.$$

The *Durbin–Watson statistic*, 'd', is

$$\text{`}d\text{'} = \sum_{i=2}^{N} (\Delta y_i - \Delta y_{i-1})^2 \left/ \Delta(\Pi y_i)^2 \sum_{i=1}^{N} \right.$$

where

$$\Delta y_i = y_{oi} - y_{ci}.$$

of Chapter 3) has often said, 'if your S value turns out to be significantly less than 1.00, you have surely done something wrong'. Chapter 3 contains an important discussion of how one should, or should not, interpret these S and the various R values.

A statistic strongly recommended by Hill and Flack (1987) is the Durbin–Watson statistic, 'd'. It is intended to reveal serial correlation between the successive y_i values. The ideal value for it is 2.00. It is now being routinely calculated in several of the more widely distributed programs for Rietveld refinement. Since the statistical errors in the intensity measurements at successive steps across a reflection profile do not depend on each other, there is no serial correlation in a statistical sense. If, however, the calculated and observed profile functions just do not match well, whether for reasons of shape or area, there will be strong serial correlation of the residuals and the D–W 'd' will be far from its ideal value of 2.00. Thus, this 'd' statistic can be a useful indicator of the quality of the fit of the calculated Bragg reflection profile functions to the actual observed profiles. Of course, it will be small (or large) if the areas, and therefore the integrated intensities which are of primary importance for crystal structure refinement, do not match. For that reason, 'd' is usually small at the beginning of a refinement series and gets progressively larger (closer to 2.00) as the refinement progresses. But if the observed and calculated profiles have different shapes, the D–W 'd' will not become 2.00 even though the areas may be perfectly matched. Consider, as a useful hypothetical example, a Gaussian observed profile and a Lorentzian calculated profile with the same area and recall that the Lorentzian has much longer 'tails' than does the Gaussian profile. The intensity at each step in the central regions of the profiles will be higher on the Gaussian profile than on the Lorentzian profile while the converse will be true for the outer regions. The result will be that 'd' calculated as shown in Table 1.3 will be significantly less than 2.00 for a non-statistical reason.

Numerical criteria are very important, but numbers are blind. It is imperative to use graphical criteria of fit, also, e.g. difference plots as well as plots of the observed and calculated patterns. That is particularly true when one is starting work on a new structure or a new sample or a new set of data for the 'same' structure. Such plots often give immediate clues to the source of the problems one is having in getting the refinement started well. Gross errors (e.g. in scale factor, lattice parameters, zero offset, wrong structure, strong phase contamination, etc.) are usually immediately obvious in the plots but not in the tables of numbers that are output from the refinements. (See Section 1.7.)

R_{wp} values can be seriously inflated by things which, one can see at once in a difference plot, do not arise from a poor structural model. An example is shown in Fig. 1.7. We see immediately that the strong line in the difference plot at 44° is probably not from the main phase and that it is strong enough to inflate the R_{wp} value significantly. In fact, that line arose from the aluminum sample holder which was improperly placed. The difference plot made it very easy to note the presence of a 'second phase'.

Conversely, R_{wp} values can be misleadingly small if the refined background is high: it is easier to get a good fit to a slowly varying background than to sets of Bragg reflection profiles.

Other criteria, both numerical and graphical, are still needed. One that may be useful is a weighted difference plot. Such plots have been used by only a few workers and they have used different weighting schemes. A scheme that accentuates angular trends and gives equal importance to the misfits in low and high intensity regions is a plot of the fractional difference at each point divided by the standard deviation, σ_i, of the observed intensity at that

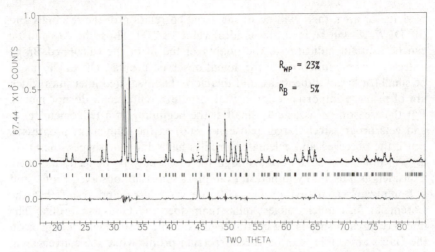

Fig. 1.7 Rietveld refinement plot for hydroxyapatite showing how the difference plot makes obvious the presence of Bragg reflections not belonging to the phase(s) modelled.

point (normally based only on counting statistics):

$$\Delta y_i(\text{weighted}) = [(y_{oi} - y_{ci})/y_{oi}]/\sigma_i. \tag{1.13}$$

Figure 1.8 shows Rietveld refinement plots for a 'hard magnet' material, $Nd_2Fe_{14}B$. The unweighted difference plot in Fig. 1.8(a) shows nothing remarkable. The weighted difference plot in Fig. 1.8(b) shows that there are, in fact, some trends of the difference with angle that are not modelled as well as they might be, even though the actual differences are not serious at any point. (The horizontal lines in the difference-plot region are placed at $+1$ and -1, i.e. at the levels where the difference would be just one σ_i at each point.)

In concluding this section, we may say that numerical criteria are needed for assessment of the fit in quantitative detail while graphical criteria are needed for a global view, easy noting of gross errors and omissions, and identification of ways in which the model may be in error. This point is further discussed in Section 1.7.

1.5 Precision and accuracy

The most often cited indicator of precision in Rietveld refinements is the estimated standard deviation (e.s.d.), calculated as σ_j in the jth parameter

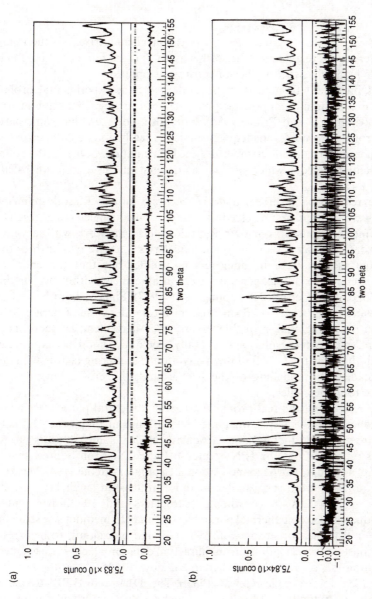

Fig. 1.8 Rietveld refinement plots (neutron data) for a Nd–Fe–B material, with difference plotted (a) without weighting and (b) with weighting.

in the following way:

$$\sigma_j = \left[M_{jj}^{-1} \frac{\sum w_i(y_i(\mathrm{o}) - y_i(\mathrm{c}))^2}{N - P + C} \right]^{1/2} \tag{1.14}$$

where M_{jj}^{-1} is the diagonal element of the inverse matrix, N is the number of observations (e.g. the number of y_i's used), P is the number of parameters adjusted, and C is the number of constraints applied.

It is important to note that the e.s.d. is *not* the experimental probable error; it is the minimum possible probable error arising from random errors alone (Chapter 3 and Prince 1985). Note also, however, that some part of the error due to model inadequacy, which is really a systematic error, may masquerade as random error in the e.s.d. calculation. Some of the sources of systematic errors which are *not* considered in the e.s.d.'s are listed in Table 1.4.

The real probable errors and the accuracy of the Rietveld-refinement results are perhaps best assessed by comparison to single-crystal results for the same materials. In our own Rietveld refinement work with laboratory X-ray data, we have generally found that the e.s.d.'s were two or three times as large as they were in single crystal studies but that the coordinate parameters, for example, generally were in agreement within two or three, often one, combined e.s.d.'s (Young *et al.* 1977). Since it is to be expected that the e.s.d.'s, being based on random errors alone (see Chapter 3), are smaller than the real probable errors, that agreement is good indeed. Thermal-parameter results have generally been not so good, perhaps because of such things as the fall-off of the X-ray atomic scattering factors with angle and inadequate modelling of the background, the profile functions, and surface roughness effects.

The e.s.d.'s and the real probable errors in structural parameters tend to be smaller, than the above, when high resolution neutron powder diffraction data are used. This first became particularly evident some ten years ago with results reported from the IPNS (Intense Pulsed Neutron Source, thus the data are TOF data) at Argonne National Labotatory in the USA. The e.s.d.'s being reported were very similar to those from good single-crystal refinements, e.g. 0.002 Å in coordinate parameters and 10–20 per cent for individual anisotropic thermal parameters in small to medium size structures. (See, for example, Jorgenson *et al.* 1981.) Similar precision is to be expected from constant wavelength neutron diffraction data for which the instrumental resolution is 2 or 3×10^{-3}, or better, e.g. now 5×10^{-4}, for the minimum $\Delta d/d$. (See, for example, Hewat 1985.) The D1A and D2B instruments, respectively, at the ILL in France have such resolutions, as will the HRNPD instrument now coming on line at the Brookhaven National Laboratory in the USA.

Table 1.4 Some sources of systematic errors in Rietveld refinement. (Context: primarily laboratory X-ray diffractometry.)

Preferred orientation
Background
Anisotropic reflection-profile broadening
Profile shapes
Absorption (differs with geometry)
Specimen displacement
Specimen transparency
Extinction
2θ-Zero error
Graininess (too few crystallites diffracting)
Incident beam instability
Instrument electrical or mechanical instability

A particularly convincing example of probable accuracy, as well as precision, obtained with TOF neutron data is given in Chapter 11. Anisotropic thermal parameters as well as coordinate parameters for benzene at low temperatures are shown to compare 'very favourably' with single crystal results. In lines 3 and 4 of Table 11.4 in Chapter 11, one sees that the agreement is generally within 2 e.s.d. (combined). Again, as for the X-ray work mentioned above, the e.s.d.'s in the powder work are 2–3 times larger than those in the single crystal work. The small scatter in the data points in Fig. 1.6, each pair based on a different set of data of Rietveld refinement, is also strong evidence of similar high precision in Rietveld refinement results from fixed-wavelength powder diffraction neutron data.

Another recent example of precision in Rietveld refinement with neutron TOF data is provided by the work of Jorgensen *et al.* (1987) on the famous '123' high temperature superconductor. They followed the shift of one of the 7 oxygen atoms (atom O(1) in Fig. 12.5) from an initially nearly full site at $0, \frac{1}{2}, 0$ to an initially vacant site at $\frac{1}{2}, 0, 0$, as a function of preparation temperature. With increasing preparation temperature, the occupanies of the two sites eventually became equal, during which process the orthorhombic-to-tetragonal phase transition took place (the two sites are crystallographically equivalent in the tetragonal phase) and the compound was no longer superconducting. At the same time, the total oxygen content of these two sites decreased with increasing preparation temperature and the combination of disordering and oxygen loss caused the material to lose its high-T_c superconducting properties when the total content of the two sites was about 0.5. Figure 1.5 shows the relevant plots of site-occupancy vs preparation-temperature. That excellent precision was achieved is attested to by the fit of the experimental points to a smooth curve. It seems to be about 1 per cent, which is about the same as the e.s.d.'s obtained for these parameters and shown in the plots by vertical error bars.

A notably comprehensive and balanced review of the 'precision and accuracy of crystal structure refinements from powder data' has been provided by Taylor (1985). He considered both neutron and X-ray powder diffraction studies. Most of the works he cited used the Rietveld refinement method, but a few used an integrated intensity method. His overall, cautiously understated conclusion is: 'It is shown that powder methods, especially the Rietveld method, have performed well in a variety of applications.'

Hill and Madsen (1986) have made some interesting direct comparisons of the calculated e.s.d.'s with experimental reproducibility assessed by replication of the experiment five times. The sample material was α-Al_2O_3. Not unexpectedly, by progressively reducing the data quality they were able to increase the calculated e.s.d.'s until they were as large as the experimentally assessed reproducibility. That is to say, until the uncertainty (e.s.d.) due to random errors (counting statistics) was the dominant component of the total uncertainty arising from all sources. In the process, they discovered that the lattice parameters are in a class by themselves (among the various usually refined parameters) in that the calculated e.s.d.'s (based on eqn (1.14)) were very much smaller in comparison to the actual reproducibility than was true for the other parameters. This result is shown graphically in Fig. 1.9, taken from their work. The counting time was 5 s per step and the step width used, in $^\circ 2\theta$, is the abscissa. The data quality was progressively reduced by increasing the step width while a fixed counting time was maintained. Further discussion of this and related work is given in Chapter 5.

1.6 Some popular computer programs for Rietveld refinement

Almost as rapidly as the applications of the Rietveld method have proliferated, so have the versions of computer programs for carrying out the refinements. Although vastly modified and extended during several generations of evolution, many and perhaps most are recognizable descendants of Rietveld's original program (Rietveld 1969) for which he freely distributed the source code. While most of those versions have been prepared only for local use, a few have been deliberately prepared and documented for general use. Recently, some versions have become available which run on 'PCs', i.e. personal computers. A report to the IUCr Commission on Powder Diffraction has recently appeared (Smith and Gorter 1991) which lists information about a large number of available programs for powder diffraction analyses, including more than a dozen for Rietveld refinement.

We list here a few powerful programs which are currently well known and

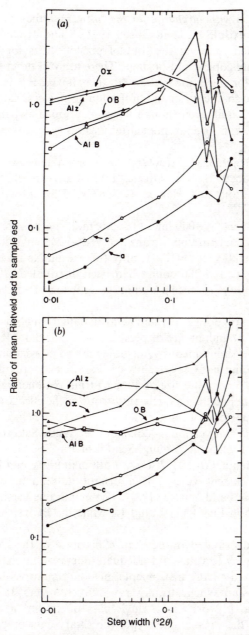

Fig. 1.9 Variation in the average ratios e.s.d.$_{exp}$ to e.s.d.$_{14}$ for six Rietveld refinement parameters in Al_2O_3, where e.s.d.$_{14}$ was calculated as in eqn (1.14) and the e.s.d.$_{exp}$ values were estimated experimentally from comparison of the results from five different data sets independently collected from the same source material at (a) 5 s step^{-1} and (b) 0.05 s step^{-1} (From Hill and Madsen 1986.)

widely distributed *without charge for the software*. Other good programs in this category do exist (Smith and Gorter 1991). Some form of reimbursement may be requested to cover the 'out of pocket' costs for things such as materials, making copies, and postage. There are still other Rietveld refine- ment programs for which the authors (or owners), having departed from the spirit of scientific co-operation and professionalism, assess substantial charges for the software, itself, even though it was developed as part of a scientific program rather than for entrepreneurial reasons.

(i) GSAS—Generalized Structure Analysis System. Authors: Allen C. Larson and Robert B. von Dreele, LANSCE, MS-H805, Los Alamos National Laboratory Los Alamos, NM 87545 USA. Obtainable from the authors.

This is a massive ($>90\,000$ lines of code), multi-purpose program package with excellent documentation. It runs in a VAX-VMS environment and, now (1992), also in UNIX. As of 1991, only the executable code was generally distributed. More than 200 copies had been distributed as of this writing and the popularity of the program is growing. It is also constantly upgraded.

GSAS works with both X-ray and neutron data from both angle dispersive (fixed wavelength) and energy dispersive (e.g. TOF) data. It can work with multiple data sets simultaneously, including mixed powder and single-crystal data. Both soft and hard constraints can be used. It provides for separate determination of crystallite size and microstrain parameters on the basis of order dependence and also models anisotropy in the effects. GSAS was used for the work reported in Chapter 12.

(ii) DBWS (current version: 9006). Authors: Wiles, Sakthivel, and Young. Available from Prof. R. A. Young, School of Physics, Georgia Institute of Technology, Atlanta, GA 30332, USA. Both mainframe and PC versions are available. The current version is a distant descendant of the program described by Wiles and Young (1981). The mainframe version was used for the work shown in Figs 1.1, 1.7, and 1.8 while the PC version was used for Fig. 1.10.

This widely distributed (more than 400 copies of the various versions) program has many features and adequate documentation. It accepts data from various fixed-energy instruments including multiple-detector neutron diffractometers and synchrotron–X-ray diffractometers. It does not work with TOF data. It works in a single-pass operation (as do GSAS and RIETAN) with the various data types (necessary data preparation is done as a part of the single pass) and all space groups. It is portable (fully ANSII 77, except for the plot routines). The source code is distributed. Users have installed the program on CDC-Cyber, IBM, and several other mainframes plus various VAX-VMS systems. Many user-modified versions are in use in various locations. It is relatively user-friendly. A special feature is the option of using a Thompson *et al.* (1987) pseudo-Voigt function as modified by Young and Desai (1989), to permit separation of crystallite-size broadening effects and microstrain broadening effects on the basis of order dependence, while permitting each effect to have both a Lorentzian and a Gaussian component.

(iii) RIETAN—Author: F. IZUMI, Nat'l Inst. for Research in Inorganic Materials, 1-1 Namiki, Tsukuba-shi, Ibariki 305, Japan. Available from: Dr Izumi.

RIETAN has become a *de facto* standard in Japan. As of late 1990, about 80 copies had been distributed in Japan and abroad. This program is part of FAT-RIETAN—an integrated system of RIETAN and related crystallographic programs. It works with TOF neutron data, and with fixed energy neutron and X-ray data including synchrotron X-ray data. Both RIETAN and FAT-RIETAN are described at some length in Chapter 13.

RIETAN has all of the usual features, including various profile functions. Both soft and hard constraints are implemented. It offers the user a choice of Gauss–Newton, modified Marquardt, and Conjugate Direction least-squares algorithms; the choice may be made anew, under programmatic control, at the start of each refinement cycle. This choice could be very helpful in assessing the existence of, and avoiding, false minima. The program can do joint refinement with X-ray and neutron data under non-linear constraints. One version of it can refine incommensurate structures as well as super-structures. It is described as a user friendly interactive system using menus, icons, and a mouse. It operates under UNIX V on a workstation (CPU: 68020, coprocessor: 68881). The program package includes ORFFE, ORTEP, Fourier and D syntheses.

(iv) 'XRS-82; The X-ray Rietveld System'. Ch. Baerlocher, Institut fuer Kristallograhie und Petrographie, ETH, Zurich, Switzerland. Avaiable from: Dr Baerlocher.

XRS-82 is described as being a complete collection of programs for Rietveld refinement and related calculations. The main programs are based on the well known X-ray System 'X-RAY 72' by Stewart. Written in FORTRAN, XRS-82 has been successfully used by a number of workers at different institutions. It is a particular favourite of people working on zeolites. It features a learned-profile-function option and sophisticated use of constraints and restraints, including various items of stereochemical knowledge. These features of the program, and their applications, are discussed in Chapter 10.

(v) Another distant descendant of the DBW2.9 program described by Wiles and Young (1981) is LHPM by Hill and Howard (1986), which has also enjoyed some general distribution. Of particular interest is a further descendant of LHPM, the QPDA program (Madsen and Hill 1990) specifically tailored for quantitative phase analysis via Rietveld refinement (appendix (5.A) to Chapter 5).

1.7   Refinement strategy

1.7.1   *Need*

A distinct strategy for refinement can save much time and disappointment. It is particularly important to think strategically when one is starting work on a new structure or with a new set of data, a different specimen of the same material, or a different kind of data, e.g. neutron vs X-ray. The following suggestions apply to Rietveld refinements with fixed-wavelength X-ray or neutron data. They will probably be helpful, but they should not be construed as being the only possible way to proceed. Many variants are possible.

1.7.2   *Model selection*

At the outset, one must have the correct space group, fairly good lattice parameters, atom coordinates, and estimated N_j's and B_j's. Stoichiometric N_j values and zero B_j's will usually let one get started. Initial estimates of U, V, W are needed. For example, setting W at the measured $(FWHM)^2$ of mid-range peaks in the observed pattern and setting $U = V = 0$ is often a good start. Initial values for X and Y will also be needed if the TCH pseudo-Voigt profile function of Thompson *et al.* (1987) is used and also for Z if the modification of the TCH profile introduced by Young and Desai (1989), to allow both the Lorentzian and the Gaussian components to have both an order-dependent and an order-independent part (for crystallite size and microstrain analysis) is used. Again, it is usually acceptable to start with $X = Y = Z = 0$.

1.7.3   *Use of graphics*

It is important to have some graphics capability available from the outset with which the observed, calculated, and difference plots can be displayed. Gross errors in scale factor, in background level or shape, or in lattice parameters, for example, can be recognized at once in such plots but not so easily in the numerical output. A plainly wrong model can also be so recognized, as can strong phase contamination. To repeat an earlier phrase, 'Numerical criteria are very important, but numbers are blind; it is imperative to use graphical criteria . . . also . . .'. As the refinement progresses, graphics are invaluable for revealing various kinds of profile misfits, including inadequate calculated-tail length, uncorrected asymmetry, anisotropic profile broadening, and intrinsic inadequacy of the profile modelling function being used. The capability for enlarging the operator's choice of small sections of the plots ('zooming in') to permit detailed examination of particular local features, such as details of shapes and misfits, can be surprisingly informative, if one has not been in the habit of doing that. Examination of the detailed shapes in the difference plots, if presented in adequate size, can also be very instructive. Figure 1.10 gives some examples, based on quartz, of easily identifiable characteristic difference-plot features arising from (a) an error of 0.008 Å in the c lattice parameter while the a parameter is correct, and (b) a 0.02 degree error in the 2θ-zero parameter.

1.7.4 *Parameter turn-on sequence*

The various refinement programs have 'switches' which one can set to select which parameters or group of parameters will be refined in the next run, i.e. 'turned on'. Different types of parameters in Rietveld refinement have quite different characteristic behaviours. Premature turn-on of non-linear, unstable parameters can lead to instant catastrophe. A possible 'good' turn-on sequence is suggested in Table 1.5, along with many comments on procedure. Once gross misfits are removed, two or three refinement cycles will suffice at each of the first two steps in the sequence.

A systematic, step-by-step, turn-on sequence is sometimes the most effective tool for identifying which parameter is causing the trouble when one encounters a situation in which the refinement does not go well and it is not clear why not. Examples of such 'not going well' are the occurrence of divergence (R_{wp} increasing instead of decreasing), a singular matrix, and very unreasonable shifts, among others. In such cases, one may start with only a few of the parameters turned on and do one refinement cycle. If the problem still exists, one reduces the number of parameters that are turned on and again tests, one cycle at a time, until the problem no longer appears. Then one can start systematically turning on the next parameter in the sequence and carrying out a single refinement cycle until the problem reappears and the troublesome parameter is thereby identified.

1.7.5 *Use of the correlation matrix*

Fairly early during the process of getting a refinement to run well, advantage can be taken of the correlation matrix:

1. To recognize redundant parameters. [Example: suppose that a 5th order polynomial is used to model a smoothly, slowly varying background. Coefficients of the highest order terms become highly correlated because they are redundant. They may also suffer large shifts which are nearly mutually compensatory but can cause the refinement to diverge or 'blow up'.]

2. To recognize parameters so strongly correlated that their shifts must be damped in order for the refinement to proceed in a useful fashion.

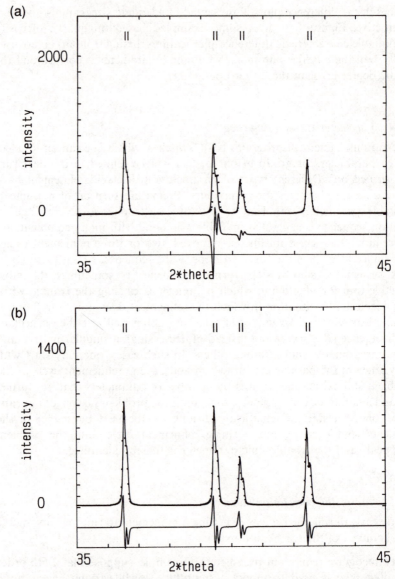

Fig. 1.10 Characteristic effects in the Rietveld refinement difference plot from specific errors in the model for α-quartz. (Figures kindly provided by Dr Ewa Sobczak, IF PAN, Warsaw.) The Bragg reflections shown are, in order of increasing angle, 11.0, 10.2, 11.1, and 20.0. (a) The c lattice parameter is in error by 0.008 Å. Note that the difference plot shows similar asymmetric features for the two reflections with l not equal to zero and no misfit for those reflections for which l is equal to zero. That is a good clue to the problem. (b) The 2θ-zero parameter is in error by 0.02°. Note that misfits shown in the difference plot are similarly shaped for all reflections, are similarly scaled to the reflection intensities, and do not depend on h, k, or l. It is, therefore, evident that the error is in a global parameter.

Table 1.5 A suggested parameter turn-on sequence. (Context: constant wavelength X-ray or neutron powder diffractometer data.)

Parameters	Linear	Stable	Comment	Sequence
Scale	Yes	Yes	Note 1	1
Specimen displacement	No	Yes	Note 2	1
Flat background	Yes	Yes		2
Lattice parameters	No	Yes	Note 3	2
More background	No	Yes(?)	Note 4	2 or 3
W	No	Poorly	Note 5	3 or 4
x, y, z	No	Fairly	Note 6	3
N's and B's	No	?	Correlated	4
U, V, etc.	No	No	Note 5	Last
Anisotropic thermal	No	No?		Last
Zero point (?)	No	Yes	Note 2	1, 4 or not

Note 1: if the scale factor is very far off or the structural model is very bad, the scale factor may get worse, e.g. smaller, during refinement because the difference between a pattern and nothing is less than the difference between two badly mismatched patterns.

Note 2: for properly aligned and mechanically stable diffractometers, the zero point error should be and remain inconsequential. In any event, it can not change from sample to sample whereas the effective specimen displacement can and does. The displacement parameter will also take up some of the effect of specimen transparency which occurs in non-infinitely absorbing specimens.

Note 3: beware lest one or more incorrect lattice parameters cause one or more calculated peaks to 'lock on' to the wrong observed peaks. The result can be a very solid false minimum. Artificially broadening the calculated profiles (temporarily) may help. A parameter for wavelength may be turned on instead of *one* of the lattice parameters if the wavelength is not as well known as are the lattice parameters.

Note 4: if more background parameters are turned on than needed to model the angular dependence of the background, the result will be high correlations and, often, large shifts that are mostly mutually compensating but may lead to erratic behaviour and failure of the refinement. The higher order ones should be turned off sequentially until the problem is corrected.

Note 5: U, V, W tend to be highly correlated. Various combinations of quite different values can lead to essentially the same profile breadth. In Chapter 3, Prince points out that the problem can be greatly ameliorated by offsetting the origin of the Caglioti *et al.* (1958) polynomial expression for the angular dependence of U, V, and W as he shows in eqn (3.12).

Note 6: graphics and the reflection indices should be used now to assess whether preferred orientation should be modelled at this point.

1.7.6 *False minima*

The global minimum is wanted, but false minima are always enticing. Rietveld refinement is more susceptible than is single-crystal refinement. Cure? None. However, one can reduce the risk by varying the starting model in various significant ways and seeing if the same minimum is reached. If so, it *may* be the global minimum. The false-minimum problem can also be alleviated by introduction of other criteria and other types of data and constraints into the refinement process. For example, a distance-least-squares procedure used with decreasing weight as the Rietveld refinement proceeds can be helpful. The use of different least-squares algorithms in the Rietveld refinement can also be helpful. It is for this reason that Izumi provides a choice of three algorithms in his program (Chapter 13).

1.7.7 *When to stop doing more and more Rietveld refinement cycles*

Unless somehow stopped, a refinement could keep on cycling forever, making insignificant adjustments to the refinable parameters. Most programs require that the operator specifies the maximum number of cycles after which the refinement is to stop automatically. For example, one might specify 2 cycles when only the scale factor and the specimen displacement parameters are being refined, but 5–10 or more when all of the parameters have been turned on and the refinement is proceeding well. It is usually good practice to set the refinement to stop after a maximum of every 10, or so, cycles. That gives the operator a needed opportunity to check the output and to decide whether more cycles are needed, or whether the refinement is essentially complete, or whether the current approach is proving to be inappropriate and should be abandoned.

It is often difficult to be sure of just when to stop, but here are some pssible criteria:

- The shifts in all parameters, x_j, are less than $\varepsilon\sigma_j$, where ε is <1. This criterion has been incorporated into the DBWS programs and one need only specify the choice of ε in the input control file. In our own work we usually use $\varepsilon = 0.3$ unless there is some reason to choose another value.

- The parameter values are simply oscillating, each parameter about some particular value. In that case, the refinement is not going to get any better with the present approach, so continuing it is just a waste of computer time and operator hope. Caution is required, though. It is easy to overlook the fact that one or two parameters are oscillating about values which, in fact, are changing enough so that significant refinement is actually proceeding, though slowly.

- Another model is available to try which may be more fruitful.

- The cost of the computer time, and operator time, to continue refinement exceeds the value of the problem.

We should note, however, that *stopping* and *finishing* are not the same thing. The refined model must make physical and chemical sense, or it is not finished . . . and even then it might be wrong!

References

Ahtee, M., Nurmela, M., Suortii, P., and Jarvinen, M. (1989). *J. Appl. Crystallogr.*, **22**, 261–68.

Baerlocher, Ch. (1984). *Proceedings of the 6th International Zeolite Conference.* Butterworth Scientific.

Buerger, M. J. (1942). *X-ray crystallography*, p. 52. Wiley, New York.

Caglioti, G., Paoletti, A., and Ricci, F. P. (1958). *Nucl. Instrum. Methods*, **35**, 223–8.

Dollase, W. A. (1986). *J. Appl. Crystallogr.*, **19**, 267–72.

Fitch, A. N., Wright, A. F., and Fender, B. (1982). *Acta Crystallogr.*, **B38**, 2546–54.

Guinier, A. (1963). *X-ray diffraction*. W. H. Freeman, San Francisco.

Hewat, A. (1985). *Chem. Scripta*, **26A**, 119–30.

Hill, R. J. and Flack, H. D. (1987). *J. Appl. Crystallogr.*, **20**, 356–61.

Hill, R. J. and Howard, C. J. (1986). Austalian Atomic Energy Commission (now ANSTO) report No. M112, Lucas Heights Research Laboratories, PMB 1, Menai, NSW 2234, Australia, 15 pp.

Hill, R. J. and Madsen, I. C. (1986). *J. Appl. Crystallogr.*, **19**, 10–18.

Iannelli, P. and Immirzi, A. (1989). *Macromolecules*, **22**, 200–5.

Immirzi, A. (1980). *Acta Crystallogr.*, **B36**, 2378–85.

Immirzi, A. and Iannelli, P. (1987). *Gazz. Chim. Ital.*, **117**, 201–6.

Jones, F. W. (1938). *Proc. Roy. Soc. Lond.*, **A166**, 16–43.

Jorgensen, J. D., Beno, M. A., Hinks, D. G., Soderholm, L., Volin, K. J., Hitterman, R. L., *et al.* (1987). *Phys. Rev.*, **B36**, 3608–16.

Jorgensen, J. D., Rotella, F. J., and Roth, W. L. (1981). *Solid State Ionics*, **5**, 143–6.

Langford, J. I. and Louër, D. (1991). *J. Appl. Crystallogr.*, **24**, 149–55.

Langford, J. I., Louër, D., Sonneveld, E. J., and Visser, J. W. (1986). *Powder Diffract.*, **1**, 211–21. [Detailed pattern decomposition with attention to profile shapes.]

Madsen, I. C. and Hill, R. J. (1990). *Powder Diffract.*, **5**, 195–9.

Pawley, G. S. (1977). *Acta Crystallogr.*, **B34**, 523–8. [Important for the use of constrained-refinement program EDNIP, which was not published until three years later, and for discussion of possibilities and problems in trying to apply Hamilton's R-ratio significance test to Rietveld refinement results.]

Pawley, G. S. (1981). *J. Appl. Crystallogr.*, **14**, 357–61. [Seminal paper on using the lattice parameters are constraints in decomposing a powder pattern into constituent Bragg reflection intensities which then 'may well be usable as the starting point for application of direct methods'; EDNIP used.]

Pawley, G. S., Mackenzie, G. A., and Dietrich, O. W. (1977). *Acta Crystallogr.*, **A33**, 142–5. [First reported use of EDNIP.]

Prince, E. (1985). In *Structure and Statistics in Crystallography* (ed. A. J. C. Wilson) pp. 95–108. Adenine Press, Guilderland, NY.

Rietveld, H. M. (1967). *Acta Crystallogr.*, **22**, 151–2. [Seminar paper no. 1.]

Rietveld, H. M. (1969). *J. Appl. Crystallogr.*, **2**, 65–71. [Seminal paper no. 2.]

Rodriguez-Carvajal, J., Martinez, J. L., and Pannetier, J. (1988). *Phys. Rev.*, **B38**, 7148–51.

Sakthivel, A., French, A. D., Eckhardt, B., and Young, R. A. (1987). In *The structures of cellulose, characterization of the solid states*, ACS Symposium Series, **340** (ed. R. Atalla) pp. 68–87. American Chemical Society, Washington, DC.

Smith, D. K. and Gorter, S. (1991). *J. Appl. Crystallogr.*, **24**, 369–402.

Stokes, A. R. (1948). *Proc. Phys. Soc. Lond.*, **61**, 382–91.

Taylor, J. C. (1985). *Austral. J. Phys.*, **35**, 519–38.

Thompson, P., Cox, D. E., and Hastings, J. M. (1987). *J. Appl. Crystallogr.*, **20**, 79–83.

Von Dreele, R. B., and Larson, A. C. (1988). *LANSCE Newsletter* No. 4, Winter 1988. Los Alamos National Laboratory, Los Alamos.

Wertheim, G. K., Butler, M. A., West, K. W., and Buchanan, D. N. E. (1974) *Rev. Sci. Instrum.*, **45**, 1369–71.

Wiles, D. B. and Young, R. A. (1981). *J. Appl. Crystallogr.*, **14**, 149–51.

Will, G. (1979). *J. Appl. Crystallogr.*, **12**, 483–5.

Will, G., Parrish, W., and Huang, T. C. (1983). *J. Appl. Crystallogr.*, **16**, 611–22.

Williams, G. H., Kwei, R. B., von Dreele, R. B., Larson, A. C., Raistrick, I. D., and Bish, D. L. (1988). *Phys. Rev.*, **B37**, 7960–2.

Wu, M. K., Asburn, J. R., Torng, C. J., Hor, P. H., Meng, R. L., Gao, L. *et al.* (1987). *Phys. Rev. Lett.*, **58**, 908–12.

Young, R. A. and Desai, P. (1989). *Arch. Nauk Mater.*, **10**, 71–90.

Young, R. A. and Sakthivel, A. (1988). *J. Appl. Crystallogr.*, **21**, 416–25.

Young, R. A., Mackie, P. E., and von Dreele, R. B. (1977). *J. Appl. Crystallogr.*, **10**, 262–9.

Young, R. A. and Wiles, D. B. (1982). *J. Appl. Crystallogr.*, **15**, 430–8.

2

The early days: a retrospective view

Hugo M. Rietveld

2.1 Introduction

As a Ph.D. student at the University of Western Australia between 1961 and 1964, I became thoroughly acquainted with X-ray and neutron diffraction techniques through experiments being conducted at the HIFAR in Lucas Heights (NSW). The emphasis was on single crystal diffraction since, even then, the powder method was regarded to be inferior, particularly for structure refinement. During that period the computer entered the scientific field and long, tedious structure factor and density calculations could be obtained more or less instantly. First with the 'Mercury' computer at the University of Oxford and the 'Silliac' computer at the University of Sydney, and later the IBM 1620 at the Physics Department of the University of Western Australia, computers became an integral part of my crystallography work. They were, incidentally, also important for my later work.

2.2 Previous efforts

After obtaining my Ph.D. degree in 1964 (Rietveld 1963), I joined the neutron diffraction group of the Reactor Centrum Nederland (now Netherlands Energy Research Foundation ECN) in The Netherlands. This group had only just been established and was principally engaged in the construction of a neutron powder diffractometer. The emphasis here was mainly on powder diffraction techniques, because it was apparent that no large, single crystals could be grown of the materials that were then of interest. The first crystal structures to be determined were rather simple and of a high symmetry, with the result that the peaks were well resolved and integrated intensities could easily be obtained for further refinement. However, with compounds more complex and of lower symmetry, the overlap of peaks became so severe that separating them became practically impossible. In an effort to overcome this problem, the resolution of the diffractometer was significantly increased by using a wavelength of 2.6 Å and by eliminating the higher order wavelength with a filter of pyrolytic graphite (Loopstra 1966). This proved to be of appreciable value, especially for structure determination. For structure refinement, however, the increase of resolution certainly resulted in a better defined pattern, but often not to such an extent that the peaks were completely resolved. The solution then was to refine the structure by using not only single Bragg reflection intensities as data, but also groups of overlapping intensities (Rietveld 1966a). This worked well, but the fact remained that all extra information contained in the profile of these overlapping peaks was lost. The following step was to separate the overlapping peaks by trying to fit Gaussian peaks using least squares procedures. This method also had its limitations, however, and did not work for severe overlap.

2.3   The use of step-wise intensities

Before the advent of computers, data reduction was a must in crystallography in order to be able to handle a relatively complex structure. Integrated intensities were therefore the smallest data elements one could work with practically. To consider using the individual intensities, y_i, constituting a step-scanned diffraction peak as data was completely unrealistic. With the experience of using computers for single crystal structure refinements and having seen their enormous capacity for handling large amounts of data, I saw that the spectre of increasing the number of data by a factor of ten by using the individual intensities, y_i, instead of the integrated intensities constituted no real barrier. In the first refinement program, the intensities y_i were corrected for background and were read in together with the value of the relative contributions each constituent peak made, i.e. the value of $w_{i,k}$ in the expression $y_i = \sum w_{i,k} S_k^2$ (Rietveld 1967), where S is the structure factor.

These values were calculated from the unit-cell dimensions and the wavelength, and zeropoint and halfwidth values measured directly from the diagram. Also, for resolved peaks the integrated values were used rather than the y_i intensities, because the Gaussian peak shape did not fit well at lower angles. Later, a correction for this asymmetry was introduced. The non-refinement of the reflection-profile parameters can be explained by the fact that the then available computer, the Electrologica X1, was not powerful enough to solve a least squares problem of more than a very limited number of parameters. With the arrival of the larger Electrologica X8 computer, the program was rewritten to include the capability of refining structure as well as profile parameters (Rietveld 1969b). Twenty-seven copies of this program, written in Algol (Rietveld 1969*a*) and later, in 1972, in FORTRAN IV, were distributed to institutes all over the world and, as is noted in Chapter 1, this has greatly contributed to the acceptance of the method.

2.4 The acceptance of the method

The method was first reported at the Seventh Congress of the IUCr in Moscow in 1966 (Rietveld 1966*b*). The response was slight, or, rather, non-existent, and it was not until the full implementation of the method was published (Rietveld 1969*b*), that reactions came. At this time, the method was mainly used to refine structures from data obtained by fixed wavelength neutron diffraction; a total of 172 structures were refined in this way before 1977 (Cheetham and Taylor 1977). In the previously mentioned paper (Rietveld 1969*b*), it had been suggested that the method could also be applied to X-ray data, but it was not until 1977 (Malmros and Thomas 1977; Young *et al.* 1977; Khattak and Cox 1977) that the method became generally accepted for X-ray as well as neutron powder diffraction, first with fixed wavelength and then also with fixed angle (energy dispersive data). This is reflected in an increasing number of citations to the original papers (Rietveld 1967 and 1969*b*) as published in the Science Citation Index. Figure 2.1 shows the number of citations between the years 1967 and 1988.

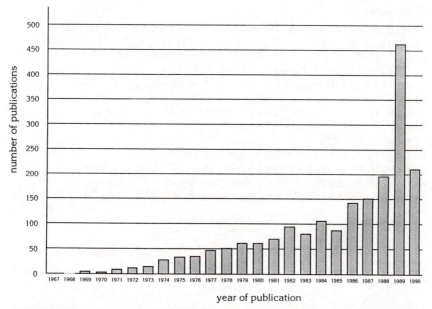

Fig. 2.1   Number of publications, in the Science Citation Index, citing as reference the original papers (Rietveld 1967, 1969*a*, 1969*b*) on the Rietveld method or having the name Rietveld in their title.

In the period January 1987 to May 1989 a total of 341 papers were published with reference to or using the Rietveld method, of which nearly half used neutron diffraction. The review article of Albinati and Willis (1982) gives a good impression of the state of the method at that moment. Many more papers on the method have appeared since, often with unexpected applications.

2.5 Conclusion

It has been most gratifying for me to experience how the Rietveld method has contributed to a renewed interest in powder diffraction techniques, even to the extent that in some applications it replaces single crystal techniques. The method is proven to be sound and has given results at least as good as single crystal data. The possible under-estimation of the actual probable errors by the calculated estimated standard deviations (see Chapters 3 and 5), may serve as a reminder to all users that the method is not to be treated as a black box. One must be continually aware of the limitations, not only of this method, but in general of all least squares methods. In this respect I fully agree with Prince (1981) who states that 'If the fit of the assumed model is not adequate, the precision and accuracy of the parameters cannot be validly assessed by statistical methods'.

References

Albinati, A. and Willis, B. T. M. (1982). *J. Appl. Crystallogr.*, **15**, 361–74.
Cheetham, A. K. and Taylor, J. C. (1977). *J. Solid State Chem.*, **21**, 253–375.
Khattak, C. P. and Cox, D. E. (1977). *J. Appl. Crystallogr.*, **10**, 405–11.
Loopstra, B. O. (1966). *Nucl. Instrum. Methods*, **44**, 181–7.
Malmros, G. and Thomas, J. O. (1977). *J. Appl. Crystallogr.*, **10**, 7–11.
Prince, E. (1981). *J. Appl. Crystallogr.*, **14**, 157–9.
Rietveld, H. M. (1963). PhD Thesis, University of Western Australia.
Rietveld, H. M. (1966a). *Acta Crystallogr.*, **20**, 508.
Rietveld, H. M. (1966b). *Acta Crystallogr.*, **21**, A228.
Rietveld, H. M. (1967). *Acta Crystallogr.*, **22**, 151–2.
Rietveld, H. M. (1969a). Research Report RCN-104. Reactor Centrum Nederland.
Rietveld, H. M. (1969b). *J. Appl. Crystallogr.*, **2**, 65–71.
Young, R. A., Mackie, P. E., and Von Dreele, R. B. (1977). *J. Appl. Crystallogr.*, **10**, 262–9.

3

Mathematical aspects of Rietveld refinement

Edward Prince

3.1   The method of least squares

The method of least squares is a powerful technique for estimating the values of the adjustable parameters in a model, $\mathbf{M}(\mathbf{x})$, that predicts the values of a set of observable quantities, $\mathbf{y}$. One seeks the minimum, as a function of $\mathbf{x}$, of the quadratic form

$$Q = [\mathbf{y} - \mathbf{M}(\mathbf{x})]^T \mathbf{W}[\mathbf{y} - \mathbf{M}(\mathbf{x})], \tag{3.1}$$

where $\mathbf{W}$ is a weight matrix that must be positive definite. The individual observations, y_i, are assumed to be drawn at random from a population whose mean is $M_i(\mathbf{x})$ when $\mathbf{x}$ has its unknown (and in principle unknowable) 'correct' value. In other words, $\langle \mathbf{y} \rangle = \mathbf{M}(\mathbf{x})$. (For definitions of statistical terms, see Schwarzenbach *et al.* 1989.) If the model is linear, so that $\mathbf{y} = \mathbf{A}\mathbf{x}$, then $\hat{\mathbf{x}}$, the set of parameters that minimizes Q, is

$$\hat{\mathbf{x}} = (\mathbf{A}^T \mathbf{W} \mathbf{A})^{-1} \mathbf{A}^T \mathbf{W} \mathbf{y}, \tag{3.2}$$

which is known as the *least squares estimate*.

An estimate, because it is a function of random variables, is itself a random variable drawn from a population with a mean and a variance. If the mean of this population is equal to the 'correct' value, if $\langle \hat{\mathbf{x}} \rangle = \mathbf{x}_c$, then the estimate is said to be *unbiased*. For the least squares estimate

$$\langle \hat{\mathbf{x}} \rangle = \langle (\mathbf{A}^T \mathbf{W} \mathbf{A})^{-1} \mathbf{A}^T \mathbf{W} \mathbf{y} \rangle = (\mathbf{A}^T \mathbf{W} \mathbf{A})^{-1} \mathbf{A}^T \mathbf{W} \langle \mathbf{y} \rangle$$

$$= (\mathbf{A}^T \mathbf{W} \mathbf{A})^{-1} \mathbf{A}^T \mathbf{W} \mathbf{A} \mathbf{x}_c = \mathbf{x}_c, \tag{3.3}$$

so that the least squares estimate is unbiased irrespective of the choice of $\mathbf{W}$.

If, however, the joint probability density function (pdf) of the populations from which the observations, y_i, are drawn has a variance–covariance matrix $\mathbf{V_y}$, and $\mathbf{W} = \mathbf{V_y^{-1}}$, then the variances of the population distributions for the elements of $\hat{\mathbf{x}}$ can be shown (see Prince 1985) to have the lowest values that they can have for any choice of $\mathbf{W}$. The particular least squares estimate

$$\hat{\mathbf{x}} = (\mathbf{A}^T \mathbf{V_y^{-1}} \mathbf{A}) \mathbf{A}^T \mathbf{V_y^{-1}} \mathbf{y} \tag{3.4}$$

is said to be the *best linear unbiased estimate* of $\mathbf{x}$.

The models for many important phenomena, including X-ray and neutron diffraction, are non-linear. The usual procedure for applying the method of least squares to a non-linear model is to find, by iterative, numerical methods, a set of parameter values, $\mathbf{x}'$, close enough to a point at which the gradient of Q vanishes for the approximation

$$M_i(\mathbf{x}) = M_i(\mathbf{x}') + \sum_{j=1}^{p} (x_j - x_j') \, \partial M_i(\mathbf{x}')/\partial x_j \tag{3.5}$$

to be a good one. A_{ij} is then set equal to $\partial M_i(\mathbf{x}')/\partial x_j$, and the least squares estimate is

$$\hat{\mathbf{x}} = \mathbf{x}' + (\mathbf{A}^T \mathbf{W} \mathbf{A})^{-1} \mathbf{A}^T \mathbf{W} [\mathbf{y} - \mathbf{M}(\mathbf{x}')]. \tag{3.6}$$

Because the point at which the approximation in eqn (3.5) is most likely to be valid is $\mathbf{x}' = \hat{\mathbf{x}}$, it is customary to iterate to full convergence. It is important, however, to distinguish between $\mathbf{x}'$, which is a basically arbitrary, displaced origin, and $\hat{\mathbf{x}}$, which is a random variable drawn from a population whose mean is $\mathbf{x}'$.

3.2   The Rietveld model

In the Rietveld method the observations are the raw data, the numbers of X-ray photons or neutrons counted at a point in a powder diffraction pattern. The model is

$$M(S_i, \mathbf{x}) = b(S_i, \mathbf{x_b}) + \sum_{k=k_1}^{k_2} I_k(\mathbf{x_s}) \phi(S_i - S_k, \mathbf{x_p}), \tag{3.7}$$

where $\mathbf{x_b}$, $\mathbf{x_s}$, and $\mathbf{x_p}$ are background, structure, and peakshape parameters, respectively, and $S_i = 2 \sin \theta_i / \lambda_i$. Either θ_i (angle dispersive diffraction) or λ_i (energy dispersive diffraction) may be the controlled, experimental variable. The term $b(S_i, \mathbf{x_b})$ is a background function, $I_k(\mathbf{x_s})$ is the integrated intensity of the kth Bragg reflection, $\phi(S_i - S_k, \mathbf{x_p})$ is a peakshape ('profile') function (see Chapters 1, 7, and 14) normalized so that the sum over the range of peak is 1, and the sum is over all reflections that can contribute to the intensity at S_i. A central part of Rietveld's (1969) contribution was to recognize that, if ϕ is well known, the information contained in overlapping peaks is not entirely lost unless they exactly coincide.

3.2.1 *Background*

Background is usually a smooth function that varies much less rapidly with S than does the diffraction pattern. It may, however, contain diffraction patterns of additional phases with known structure. (See Chapter 6.) The smooth part may be represented by a polynomial, or a sum of polynomials, each term of which has a linear scale factor. The impurity phase part is best represented by a pattern measured from a pure sample of that phase, or calculated from its structure as if it had been measured, under the conditions of the current experiment, multiplied by a scale factor. The simultaneous refinement of the structures of two or more phases of unknown structure is controversial.

3.2.2 *Peak shapes*

Each of the different experimental techniques for collecting data, constant wavelength and time-of-flight for neutrons and characteristic line source and synchrotron source of X-rays, has its own most appropriate peakshape function (Chapters 7, 11, 13, and 14; Young and Wiles 1982). Because peak shapes tend to be the result of a number of different contributions, and because the convolution of multiple resolution functions tends toward a Gaussian function (the central limit theorem), the underlying shape tends to be Gaussian, but it may have a convolution with some other curve to allow for the particular effects in a specific technique. The Rietveld method was developed originally for the constant wavelength, medium-resolution, neutron case, in which, at least with well crystallized samples that are free from crystallite size or inhomogeneous strain effects, the Gaussian is an adequate representation of the peak shape. X-ray peaks tend to have longer tails than a Gaussian, and may be somewhat asymmetric. The peaks in a time-of-flight neutron pattern are always markedly asymmetric and thus require a more complex function (Chapters 11 and 14). Because of the curvature of a Debye–Scherrer cone and the axial divergence in the diffractometer, low angle peaks are asymmetric even in constant wavelength geometry. Various schemes have been used (Rietveld 1969; Howard 1982; Prince 1983) to allow for this effect. All of the peakshape functions vary in width as a function of S, and some vary in shape as well. For the constant wavelength case the UVW formula of Caglioti *et al.* (1958), which gives the square of the full width at half maximum as a quadratic in tan θ, has served well for low and medium resolution neutron powder diffraction patterns. It has been widely employed as a useful approximation for X-ray and high-resolution neutron powder diffraction patterns. The effect of the peakshape function on the calculated intensity depends on the calculated position of the peak, which depends in turn on the cell constants, so they, or, in actual practice, elements of the reciprocal space metric tensor, are also adjustable, peakshape parameters.

3.2.3 Weights

Equation (3.4) indicates that ideally, the weight matrix in the least squares estimate should be the inverse of the variance–covariance matrix for the observations. In the Rietveld method the observations are raw counts for which, if the apparatus works properly, the statistical fluctuation term in any one should be unaffected by any other. The observations are therefore statistically independent, and the weight matrix is diagonal, with $W_{ii} = 1/\sigma_i^2$, where σ_i^2 is the variance of the ith observation. The statistical variation of the counting of random events has the Poisson distribution, for which the probability of observing k counts is

$$P(k) = \zeta^k \exp(-\zeta)/k!, \tag{3.8}$$

in which ζ is a parameter that represents the 'true' average count rate. This distribution has both its mean and its variance equal to ζ, so the ideal weight of the ith term in the sum of squares should be $1/\zeta_i$, but ζ_i is not known a priori. The fact that the mean of the distribution is ζ implies that the average of many repeated measurements of k will be close to ζ, but, for a single observation, k is *not* an unbiased estimate of ζ. That this must be true may be seen by considering the case where $k = 0$. The parameter ζ may not be negative, but it may have a positive value over a range and still give a significant probability of observing $k = 0$. Bayes's theorem (Schwarzenbach *et al.* 1989) may be used to construct an unbiased estimate of ζ, but it depends on an assumption as to the 'prior' distribution in the absence of previous information. Rainwater and Wu (1947; also Nicholson 1966) give an analysis that essentially assumes that the prior distribution is uniform over the entire range of possible values and conclude that $\hat{\zeta} = k + 1$ is an unbiased estimate of ζ. Box and Tiao (1973), however, argue that a prior distribution that is 'locally noninformative' should vary as $\zeta^{-1/2}$, and conclude that $\hat{\zeta} = k + 1/2$ is a proper, unbiased estimate. Because it is difficult to imagine that the prior would be an increasing function of ζ, we may assume that the Rainwater and Wu result is a worst case, and that an unbiased estimate of λ is $k + \delta$, where δ is a positive number ≤ 1. The important point is that δ is independent of k, and the relative bias gets small as k increases.

Another reason, perhaps, for not setting the weight equal to $1/y$ is that it gives a greater weight to points that represent negative fluctuations than to ones that represent positive fluctuations, thereby applying a negative bias to all intensities. The results given in the previous paragraph, however, suggest that the effect is to lower the entire calculated pattern by a constant of order 1. It is difficult to imagine why a bias of this magnitude in the estimate of the constant term in the background function would ever be of concern in any experiment, but reference to eqn (3.7) shows that this is the only parameter that is affected in any way. It would seem, therefore, that $W_{ii} = 1/y_i$ is adequate for almost all practical purposes.

3.2.4 *Multiple data sets and other information*

Each term in a weighted sum of squares is the square of the difference between an observable quantity and the prediction of a model, expressed as a multiple of the standard deviation of the observation. There is no requirement that all of the observable quantities be measured in the same experiment, or even by the same technique. They can even be items of auxiliary information, such as ideal values of bond distances and angles (Chapter 10; Baerlocher *et al.* 1977). Two examples of effective use of combined neutron and X-ray data sets are discussed in Chapter 12.

An important particular case of multiple data sets is that of a diffractometer with multiple detectors collecting data simultaneously. In time-of-flight neutron diffractometry it is customary to use 'electronic focusing' to combine the data from several detectors in a bank into a single data set (Chapter 11). This procedure is satisfactory provided the uncertainties in the time corrections are small compared with the widths of the individual time bins. In the case of a constant wavelength diffractometer with an array of detectors in a fan arrangement, it is difficult to construct the instrument so that the angle intervals between detectors are precise integral multiples of the step size, with the result that the intensities at points on the sides of sharp peaks that are observed by two or more detectors are not exact duplicates, and it is safer to treat the data from the different detectors as independent data sets rather than to use preprocessing to merge them into a single set.

3.3 Constrained models

The integrated intensities, $I_k(\mathbf{x_s})$, in eqn (3.7) are functions of a linear scale factor and Lorentz and polarization factors that are assumed to be known with far greater precision than the measured intensities. $I_k(\mathbf{x_s})$ is also a function of crystallographic parameters, atom positions, thermal displacements, and site occupancies, which are the quantities the crystallographer wishes to determine. Although the Rietveld method makes optimum use of the information contained in a composite peak, there is inevitably less usable information than there would be in a single crystal pattern of the same phase. It is therefore even more important than it is in single crystal diffraction to incorporate in the model, by means of a system of constraints, as much as possible of what is already known. (See Chapter 10.)

To implement a constrained model, define a set of relations that specify the unconstrained parameter set, x_u, in terms of a smaller set of parameters, x_c, which may be, but need not be, a subset of the unconstrained parameter set. Then define a constraint matrix, C, by

$$C_{jk} = \partial x_{uj}/\partial x_{ck}. \tag{3.9}$$

The least squares estimate is then

$$\hat{x}_u = x'_u + C(C^T A^T W A C)^{-1} C^T A^T W[y - M(x'_u)]. \tag{3.10}$$

In crystallographic refinement the most frequently encountered constraints are those imposed by space group symmetry on the positional and thermal displacement parameters of atoms in special positions. The partial derivatives of the structure factors with respect to some independent parameter are then linear combinations of the partial derivatives with respect to all parameters that depend on it. Another situation in which constraints are important occurs when atoms of two or more chemical species occupy two or more crystallographically non-equivalent sites in an ordered, partially disordered, or fully disordered fashion. The overall stoichiometry may be well known from chemical analysis, and the total occupancy of a site is known (one of the 'species' may be a vacancy), but the distribution of the various species among the various sites must be determined. The constraints take the form (Finger and Prince 1975) of linear equations giving the average atomic scattering factor for each site in terms of the numbers of some of the kinds of atoms in some of the sites, while holding the total numbers of each kind of atom and the total occupancies of all sites constant.

Another useful type of constraint is the imposition on a cluster of atoms of a symmetry that is not reflected in the space group and/or to require it to vibrate as a rigid body (Prince and Finger 1973; Trevino et al. 1980; Prince 1982a; Fitch et al. 1986). Both of these can be accomplished by defining a special coordinate system relative to which the atom positions can be specified and the rigid body motion tensors can be defined. Three positional parameters define the origin of the special coordinate system relative to the origin of the unit cell, and three more define its orientation with respect to the crystallographic axes. (Euler angles usually can be used, but other representations are sometimes more convenient.) The expressions that relate positions in the special system to those in the unit cell in terms of rotations of the coordinate system are non-linear, as are the expressions for the thermal displacement parameters in terms of the elements of the rigid body motion tensors, **T**, **L**, and **S** (Schomaker and Trueblood 1968; Prince and Finger 1973; Prince 1982*a*), so that the elements of **C** must be recomputed in each cycle. Otherwise the procedure is straightforward and can produce physically realistic models with a large reduction in the numbers of adjustable parameters.

3.4 Refinement procedures

Crystallographers were among the first scientists to exploit the development of high speed computers to find, by iterative methods, the solution to multiparameter fitting problems. One result of the fact is that many of the crystallographic least squares programs, including Rietveld's original program, predate a large body of research by mathematicians and computer scientists on the efficient use of computers to solve such problems. Because the crystallographers were not mathematicians or computer scientists, and because the programs they were using were, for the most part, giving them satisfactory results, this research did not get incorporated in the crystallographic programs. In particular, most crystallographic programs use a procedure in which, in each iteration, a linear approximation to the model is used to construct an approximate normal equations, or Hessian, matrix for the sum-of-squares function that is then inverted to compute a new guess for the values of the parameters, a procedure that is known as the Gauss–Newton algorithm. This procedure is easy to implement, and, if the model is reasonably well constructed, and the initial guess is reasonably close to the correct structure, it will converge rapidly to a stable minimum. If either of the conditions is not met, however, the algorithm can be unstable, and expensive computer time is then spent producing nonsense. An *ad hoc* solution to this problem has been to introduce a 'damping factor' that reduces the sizes of the step in parameter space. This often helps, but it slows down the convergence of favourable cases, and it still does not guarantee stability. (See Chapter 13.)

Much better stability and, at the same time, somewhat greater speed can be obtained by incorporating two measures that together comprise a so-called 'quasi-Newton method' (Gill *et al.* 1981). First a search is performed along a line in parameter space determined from an initial approximation to the Hessian matrix, to find a point that is a better fit than the starting point. Then the gradient of the sum of squares at the initial point, the gradient at the new point and the vector between them are used to compute a correction to the Hessian, an 'update', that gives the correct gradients at both points. The line search need not find the exact minimum along the line, but insisting on a certain amount of improvement assures convergence.

Likewise, the initial approximation to the Hessian need not be the full matrix. A diagonal matrix that causes the initial line search to be in the steepest descents direction is sufficient, but experience indicates that convergence is faster if the full matrix is computed once. When the matrix has been updated, a new search direction is calculated and the process is repeated until the gradient of the sum of squares is sufficiently small.

While a quasi-Newton procedure assures that a stable minimum will be found, there is no assurance that this minimum will be the true, global minimum. The Rietveld method is particularly susceptible to being trapped in false minima, especially if the cell constants are not initially well known. This may cause peaks to be mis-indexed, and the incorrect correspondence between peaks in the observed and calculated patterns then results in a situation from which the computer cannot, by itself, escape. Frequent plots of observed and calculated patterns help to detect this condition.

Another technique that may help establish a correct initial indexing of the pattern has been described by Bartell and Caillat (1987). Both the observed and calculated patterns are first blurred by convolution with a known broadening function, which allows the computer to assign indices on the basis of intensities. Then, when better guesses of the cell constants have been found, the sharp pattern can be refined. Use of this technique resulted in a successful refinement of the previously intractable, low symmetry structure of SF_6 (Bartell *et al.* 1987).

Another effect that can cause trouble in refinement is large correlation between parameters. This is a symptom of a poor choice of adjustable parameters, so that changes in several parameters have very similar effects on the calculated intensities. The correlations between occupancy factors and temperature factors are well known. If the occupancy factors are to be refined without constraints on overall composition, it is essential for the diffraction pattern to cover as wide a range of $\sin \theta / \lambda$ as possible. In the Rietveld method a still more important set of correlations is among the peak width parameters, U, V, and W. If the formula of Caglioti *et al.* (1958),

$$H^2 = U \tan^2 \theta + V \tan \theta + W, \qquad (3.11)$$

is used in its simplest form, the correlations among the parameters can be extremely high, sometimes approaching 1.0 (a singular matrix). The refinement is unstable, and the shifts can lead to negative values of H^2 at some point in the pattern, leading in turn to a program crash on an attempt to extract the square root of a negative number and a frustrated crystallographer. The problem can be greatly ameliorated by expressing the formula in the form

$$H^2 = U'(\tan \theta - \tan \theta_0)^2 + V'(\tan \theta - \tan \theta_0) + W', \qquad (3.12)$$

where θ_0 is somewhere near the middle of the pattern.

3.5 Estimates of uncertainty

3.5.1 *Goodness of fit*

If a data set consists of independent observations, so that the weight matrix is diagonal, if the weights are the reciprocals of the variances of the populations from which the individual observations are drawn, and if the model is the unknown, true model, eqn (3.1) reduces to a sum of terms each of which has expected value 1, so that the sum of n terms is n. It can be shown (Hamilton 1964; Prince 1982a) that each parameter that is estimated reduces the sum by 1, so that the sum of n terms each of which is the weighted square of the difference between an observation and an estimated model with p parameters, the weighted sum of squared residuals, is $n - p$. The weighted pattern R index, R_{wp}, is defined by (Chapter 4; Young *et al.* 1982)

$$R_{wp} = \left\{ \sum_{i=1}^{n} w_i [y_i - M_i(\hat{x})]^2 \middle/ \sum_{i=1}^{n} w_i y_i^2 \right\}^{1/2}. \tag{3.13}$$

The numerator of the expression inside the braces is the sum of squared residuals and thus has expected value $n - p$. The expected R index, R_e, is accordingly defined as

$$R_e = \left\{ (n - p) \middle/ \sum_{i=1}^{n} w_i y_i^2 \right\}^{1/2}. \tag{3.14}$$

(Note that, strictly, this is not the expected value of $R_{wp} - R_e = \langle R_{wp}^2 \rangle^{1/2} \neq \langle R_{wp} \rangle$—but for practical purposes it is an adequate approximation.) The ratio R_{wp}/R_e, sometimes referred to as the 'χ factor' (and also as the 'goodness of fit', Chapters 1 and 5), is a measure of how well the fitted model accounts for the data.

A χ factor greater than about 1.5 is a strong indication of an inadequate model or a false minimum. A value of the χ factor less than 1.0, however, is an indicator not of an extremely high quality refinement but of a model that contains more parameters than can be justified by the quality of the data. Note also that a low value of the χ factor can be obtained by a high value of R_e, due to insufficient counting time, as well as by a low value of R_{wp}, and that a low value of R_{wp} can be obtained if there is a high background that can be fit by a relatively crude background function. These measures of goodness of fit must not substitute for scientific judgement.

3.5.2 *Uncertainties in parameters*

If a vector of random variables, x, is a linear function of another vector of random variables, y, that is $x = By$, where B is a matrix, then the variance–covariance matrix for $\mathbf{x}$ is related to that for $\mathbf{y}$ by $\mathbf{V_x} = \mathbf{BV_yB}^T$. If $\mathbf{y}$ is the set of observations and $\mathbf{x}$ is $\hat{\mathbf{x}}$, the least squares estimate given in eqn (3.4), then

$$\mathbf{V_x} = (\mathbf{A}^T\mathbf{V_y}^{-1}\mathbf{A})^{-1}. \tag{3.15}$$

The square roots of the diagonal elements of this matrix are the estimated standard deviations of the corresponding parameter estimates, a measure of the irreducible minimum in the uncertainty in the values of the parameters *if* the model is a correct representation of the data. It is customary to multiply the e.s.d.'s by the χ factor, on the assumption that the lack of fit is due entirely to a uniform over-weighting of all data points. While this practice is conservative, it should be understood that it has no basis in statistics. Defects in the model may or may not be correlated with the parameters that are included in the model, with the result that their estimated values may be more (up to the e.s.d.'s not multiplied by the χ factor) or less precise estimates of the 'true' value and may or may not be systematically biased.

3.5.3 *Comparison of different models*

It frequently happens that two or more models lead to refinements that give very similar fits to the data. It then becomes necessary to decide whether one model gives a 'significantly' better fit to the data than another. This question can be put in more quantitative terms by asking what the probability is that the observed difference between some measure of the fits of two models could occur by chance if the models were actually equally good representations of the data. For a properly weighted least squares fit of a correct model with p parameters to n data points, the probability that the sum of squared residuals, $Q(\hat{\mathbf{x}})$, will be less than a number α is given by the χ^2 cumulative distribution function[†],

$$P[Q(\hat{\mathbf{x}}) < \alpha] = [2^{\nu/2}\Gamma(\nu/2)]^{-1} \int_0^\alpha \exp(-\chi^2/2)(\chi^2)^{\nu/2-1}\,d\chi^2, \tag{3.16}$$

where $\nu = n - p$ is the number of 'degrees of freedom' of the fit and $\Gamma(x)$ is the gamma function. The expected value of the sum will be reduced by 1 for each additional parameter, even if the validity of the model as a representation of the data is not changed at all. Consider two models, an unconstrained one with p parameters and a constrained one with q parameters that are a subset of the p parameters, and let Q_u and Q_c be the sums of squared residuals for the two models. The ratio $F = [(Q_c - Q_u)/(p - q)]/[Q_u/(n - p)]$

[†] FORTRAN source code for the χ^2, F, and Student's t cumulative distribution functions is given in Prince (1982*a*).

has the F distribution, for which

$$P(F < \alpha) = \frac{\Gamma(v_1 + v_2)/2]}{\Gamma(v_1/2)\Gamma(v_2)}\left(\frac{v_1}{v_2}\right)^{v_1/2}\int_0^\alpha \frac{u}{[1 + (v_1/v_2)u]^{(v_1 + v_2)/2}}\,du, \quad (3.17)$$

where $v_1 = p - q$ and $v_2 = n - p$. If $1 - P(F < F_{obs})$, where F_{obs} is the value actually observed, is less than about 0.05, the improvement is generally considered to be 'significant'.

Starting from the F distribution, Hamilton (1964, 1965) derived a probability distribution for the ratio of the weighted R indices of two models. Because R_{wp} has the same form as R_w for single crystal studies, Hamilton's test for significance may be used for R_{wp} also. However, because the ratio tends to be close to 1.0, R_{wp} must often be calculated to many significant figures in order to perform the test. Hamilton's test and an F test are exactly equivalent.

F ratio tests are valid only if the parameter set of the constrained model is a subset of the set for the unconstrained model. Hamilton (1965) devised a procedure for comparing two different models by making one of the parameters of the unconstrained model be a linear, mixing parameter for the two models. An equivalent, and somewhat more straightforward, means for making this kind of comparison has been described by Prince (1982b). Consider two models whose calculated intensities at point S_i are $M_1(S_i)$ and $M_2(S_i)$, while the observed intensity is y_i. Let $x_i = [M_1(S_i) - M_2(S_i)]/y_i^{1/2}$, and let $z_i = \{y_i - (1/2)[M_1(S_i) + M_2(S_i)]\}/y_i^{1/2}$. Find the slope, η, of the line $z = \eta x$ that minimizes the quantity

$$f(\eta) = \sum_{i=1}^n (z_i - \eta x_i)^2. \quad (3.18)$$

This slope is

$$\hat{\eta} = \sum_{i=1}^n z_i x_i \bigg/ \sum_{i=1}^n x_i^2, \quad (3.19)$$

and its estimated variance is

$$\sigma_\eta^2 = \left(\sum_{i=1}^n z_i^2 - \eta^2 \sum_{i=1}^n x_i^2\right)\bigg/\left((n-1)\sum_{i=1}^n x_i^2\right). \quad (3.20)$$

Variable $\hat{\eta}$ will be positive if M_1 is a better model and negative if M_2 is a better model; $\hat{\eta}$ is a random variable drawn from a population with Student's t distribution, from which the probability can be calculated that its observed value would be observed if the true value of η were 0, meaning that the fits of the two models differ by no more than would be expected by chance.

It must be emphasized that all statistical measures are measures only of precision. They put a lower limit on the uncertainty of an estimated parameter, and indicate which of two models is more likely to be the correct one. For a parameter estimate, an estimate standard deviation tells nothing about bias, which can cause a precise estimate to be grossly wrong. For a comparison, the statistical test tells nothing about whether either model is correct, or whether there is still another model that fits even better.

References

Baerlocher, C., Hepp, A., and Meier, W. M. (1977). *DLS76 manual*. Institut für Kristallographie und Petrographie, ETH, Zürich.

Bartell, L. S. and Caillat, J. C. (1987). *J. Appl. Crystallogr.*, **20**, 461–6.

Bartell, L. S., Caillat, J. C., and Powell, B. M. (1987). *Science*, **236**, 1463–5.

Box, G. E. P. and Tiao, G. C. (1973). *Bayesian inference in statistical analysis*. Addison-Wesley, Reading, MA.

Caglioti, G., Paoletti, A., and Ricci, F. P. (1958). *Nucl. Instrum.*, **3**, 223–8.

Finger, L. W. and Prince, E. (1975). *A system of Fortran IV programs for crystal structure computations*. National Bureau of Standards, Washington.

Fitch, A. N., Jobic, H., and Renouprez, A. (1986). *J. Phys. Chem.*, **90**, 1311–18.

Gill, P. E., Murray, W., and Wright, M. H. (1981). *Practical optimization*. Academic Press, London.

Hamilton, W. C. (1964). *Statistics in physical science*. Ronald Press, New York.

Hamilton, W. C. (1965). *Acta Crystallogr.*, **18**, 502–10.

Howard, C. J. (1982). *J. Appl. Crystallogr.*, **15**, 615–20.

Nicholson, W. L. (1966). *Nucleonics*, **24**, 118–21.

Prince, E. (1982a). *Mathematical techniques in crystallography and materials science*. Springer, New York.

Prince, E. (1982b). *Acta Crystallogr.*, **B38**, 1099–100.

Prince, E. (1983). *J. Appl. Crystallogr.*, **16**, 508–11.

Prince, E. (1985) *Structure and statistics in crystallography* (ed. A. J. C. Wilson), pp. 95–108. Adenine Press, Guilderland, NY.

Prince, E. and Finger, L. W. (1973). *Acta Crystallogr.*, **B29**, 179–83.

Rainwater, L. J. and Wu, C. S. (1947). *Nucleonics*, **1**, No. 2, 60–9.

Rietveld, H. M. (1969). *J. Appl. Crystallogr.*, **2**, 65–71.

Schomaker, V. and Trueblood, K. N. (1968). *Acta Crystallogr.*, **B24**, 63–76.

Schwarzenbach, D., Abrahams, S. C., Flack, H. D., Gonschorek, W., Hahn, Th., and Huml, K. (1989). *Acta Crystallogr.*, **A45**, 63–75.

Trevino, S. F., Prince, E. and Hubbard, C. R. (1980). *J. Chem. Phys.*, **73**, 2996–3000.

Young, R. A. and Wiles, D. B. (1982). *J. Appl. Crystallogr.*, **15**, 430–8.

Young, R. A., Prince, E., and Sparks, R. A. (1982). *J. Appl. Crystallogr.*, **15**, 357–59.

4

The flow of radiation in a polycrystalline material

Terence M. Sabine

4.1   Introduction

The essence of the Rietveld method for the reduction of powder diffraction data is minimization, by least squares methods, of the magnitude of the difference between the observed values of the ordinates of the powder pattern, y_{iobs}, and the values of these ordinates, y_{icalc}, obtained from a model of the scattering system. The individual differences are allocated weights proportional to the reciprocal of the variance of the observed ordinate.

As is pointed out in Chapter 3, it is a consequence of the Gauss–Markov theorem that this procedure will provide the best unbiased estimates of the values of the quantities contained in the model if, and only if, the expected value of the error term is zero. To satisfy this condition the model must reflect the physical parameters of the scatterer, that is, the model must be free from systematic error. For this to be so, all processes which modify the intensities of the incident and diffracted beams during their passage through the specimen must be incorporated in the model in their correct parametric form.

It is assumed that the powder, or polycrystalline body consists of a very large number of perfect crystal blocks. In a brittle material each grain is likely to be a perfect crystal (also called a mosaic block), however in ductile materials the perfect crystal blocks will be the subgrains which are separated by small angle boundaries formed by dislocations (Read 1953). It will also be assumed that the powder pattern average of Weinstock (1944) holds, that is, there is no preferred orientation.

In this chapter the following physical processes are incorporated into the model.

4.1.1 Absorption

This process removes radiation during its passage through the specimen. The boundary conditions are the shape and size of the specimen and the configuration of the experiment. Two common configurations are a cylinder bathed in the beam (Debye–Scherrer) or a flat plate moving in a $\theta{:}2\theta$ relationship with the counter (Bragg–Brentano).

4.1.2 Multiple scattering

This occurs when a diffracted beam on the primary Debye–Scherrer cone acts as an incident beam for the generation of secondary Debye–Scherrer cones. Its effect is to remove radiation from the Bragg peaks and to redistribute it into the background. Again the boundary conditions are the shape and size of the specimen and the configuration of the experiment. From conservation of energy considerations the multiple incoherent scattering is self-compensating.

4.1.3 Extinction and micro-absorption

Extinction is the process by which the radiation reflected by a 'set of planes' is re-reflected again by a Bragg process during its passage through a crystal. The result is augmentation of the incident beam at the expense of the diffracted beam. In a random powder, extinction processes will take place within a single grain. This is usually termed primary extinction. Secondary extinction, which is the reflection by one grain by a beam diffracted by another grain will not occur unless the degree of preferred orientation is such that the specimen approaches the classic mosaic crystal.

Absorption within a single grain has a profound effect on the severity of the extinction process. This effect is different from the depletion of the diffracted beam referred to above and will be termed micro-absorption.

For historical reasons the non-kinematic effects will be expressed by the factors AME in the equation

$$y_{i\text{obs}} = AME\, y_{i\text{kin}}. \tag{4.1}$$

The ordinate $y_{i\text{kin}}$ is calculated on the basis that the integrated intensity of a reflection is equal to the intensity scattered by a unit cell, multiplied by the number of units cells in the crystal.

A is the correction factor for absorption, E is the correction factor for extinction, and M is the correction factor for multiple scattering.

4.2   The flow of radiation

Each grain of powder is bathed in radiation which enters first through the entrance surface of the specimen and then through the entrance surface of the individual grain. The diffracted beam then emerges through the exit surface of the grain and the sum of these beams through the exit surface of the specimen.

The interplay between the incident and diffracted beams is expressed by the Hamilton–Darwin equations (Darwin 1922; Hamilton 1957).

$$\partial I_i / \partial t_i = \tau I_i + \sigma I_f \tag{4.2}$$

$$\partial I_f / \partial t_f = \tau I_f + \sigma I_i \tag{4.3}$$

I_i is the intensity (particles $m^{-2}\,s^{-1}$) of the incident beam along t_i, while I_f is the intensity of the diffracted beam along t_f. The cross-section per unit volume for Bragg scattering is σ. The cross-section per unit volume for all processes which remove radiation is $-\tau$. These processes included Bragg scattering, absorption, incoherent scattering, multiple scattering, and thermal diffuse scattering.

These equations have analytic solutions only in the Laue case ($2\theta = 0$) and the Bragg Case ($2\theta = \pi$). Correction factors evaluated at these limits will be denoted by the subscripts L and B respectively.

At intermediate angles of scattering the extinction factor is given by the formula (Sabine 1988),

$$E(2\theta) = E_L \cos^2 \theta + E_B \sin^2 \theta. \tag{4.4}$$

An identical average applies for A and M.

4.3   Absorption and multiple scattering

For these processes, which involve the whole specimen, the feedback term in eqn (4.2) can be neglected and the radiation flow is then represented by

$$\partial I_i / \partial t_i = \tau I_i \tag{4.5}$$

$$\partial I_f / \partial t_f = \tau I_f + \sigma I_f. \tag{4.6}$$

The solutions to these equations are then, for the absorption correction

$$A_L = e^{-\mu D} \tag{4.7}$$

$$A_B = e^{-\mu D} \sinh(\mu D)/(\mu D). \tag{4.8}$$

The linear absorption coefficient, which is the absorption cross-action per unit volume, is μ. D is a path length parameter which depends on the shape of the specimen. The value of D is $3D/4$ for a sphere of diameter D and $8D/(3\pi)$ for a cylinder of diameter D.

A form of absorption correction appropriate to neutron diffraction experiments which is often used is that given by Hewat (1979). A numerical calculation shows that his correction is essentially identical with the formula given here.

The form of the solutions for multiple scattering are identical to those for absorption. The appropriate linear coefficient is given by

$$\mu_m = \tfrac{1}{2} N_c^2 \lambda^2 \sum_d^{\lambda < 2d} dF_d \qquad (4.9)$$

where N_c is the number of unit cells per unit volume, λ is the wavelength, F_d is the magnitude of the structure factor (including the Debye–Waller factor) for the 'reflection from planes' of spacing d. For time-of-flight experiments the summation over d must be carried out for every reflection.

In the short wavelength limit $\mu_m \to N_c \sigma_{coh}$, where σ_{coh} is the total coherent scattering cross-section per unit cell.

The effective specimen size D can be refined by least-squares methods.

4.4 Extinction

Unlike macro-absorption and multiple scattering, which are functions of the size and shape of the *specimen*, extinction is dependent on the size and shape of the *mosaic blocks*.

For the extinction process eqns (4.2) and (4.3) must be used. The length of the side of the block, which in this model is also the mean path-length for the diffracted beam in the crystal, is l.

For the Laue case, use of the substitutions $t_i = t_f = t$ and the boundary conditions

$$I_i = I_0 \text{ for } t = 0, \; I_f = 0 \text{ for } t = 0$$

lead to the solution

$$I_f = (I_0/2) \, e^{-\mu l} (1 - e^{-2\sigma l}). \qquad (4.10)$$

While for the Bragg case the substitutions $t_i = t$ and $t_f = l - t$, where l is the thickness of the block, and the boundary conditions $I_i = I_0$ for $t = 0$ and

$I_f = 0$ for $t = D$ leads to the solution

$$I_f = I_0 \sigma \sinh(al)/(a \cosh(al) - \tau \sinh(al))$$

where $a = (\tau^2 - \sigma^2)^{1/2}$.

In applying these equations it is necessary to express σ in terms of crystallographic quantities. By standard methods (Marshall and Lovesey 1971) the result (Sabine 1985).

$$\sigma(\Delta k) = Q_k \, \delta(\Delta k) \tag{4.11}$$

is obtained where $Q_k = N_c^2 \lambda^2 F^2/\sin \theta$ and here k is taken to be $2(\sin \theta)/\lambda$ (elsewhere in this book, it is taken to be $4\pi(\sin \theta/\lambda)$. N_c is the number of unit cells per unit volume, λ is the wavelength of the radiation and F is the modulus of the structure factor of the reflection under consideration. In all of the equations used in this work, F includes the temperature factor. The deviation from the exact Bragg position is Δk, which is equal to $2(\cos \theta)/\lambda$.

In calculating the extinction correction the delta function is replaced by the Lorentzian function

$$\delta(\Delta k) = \frac{TC}{1 + (\pi TC \, \Delta k)^2} \tag{4.12}$$

where T is the thickness of the crystal normal to the diffracting plane and

$$C = \tanh(\mu l/2)/(\mu l/2). \tag{4.13}$$

T is related to l by

$$T = l \sin \theta. \tag{4.14}$$

4.5 Application in Rietveld analysis

It will be assumed that the corrections have been made for macro-absorption and multiple scattering, either by the methods of this paper, or by other methods.

The extinction factor is then calculated on the basis that the intensity at the entrance surface of each grain is given by the product $I_{kin} AM$. The following relationships then apply

$$E(2\theta) = E_L \cos^2 \theta + E_B \sin^2 \theta \tag{4.15}$$

$$E_L = (1 - x/2 + x^2/4 - 5x^3/48 + 7x^4/192), \qquad x \le 1 \tag{4.16}$$

$$E_L = [2/(\pi x)]^{1/2}[1 - 1/(8x) - 3/(128x^2) - 15/(1024x^3)], \qquad x > 1 \quad (4.17)$$

$$E_B = 1/(1 + Bx)^{1/2} \tag{4.18}$$

$$x = C(N_c \lambda F l)^2$$

$$B = (1/\mu l) \exp(-\mu l)/\sinh(\mu l) \tag{4.19}$$

$$C = \tanh(\mu l/2)/(\mu l/2). \tag{4.20}$$

The quantity, l, is the mosaic block size, assuming a cube of edge l. It is equal to $3l/4$ for a sphere of diameter l and $8l/3\pi$ for a cylinder of diameter l. Anisotropic mosaic blocks can be catered for by redefinition of l (Coppens and Hamilton 1970). This is not done here. While micro-absorption (except in pathological cases) can be ignored for neutron diffraction, it is not negligible for X-rays and must be included when simultaneous refinement of neutron and X-ray data is attempted.

The parameter l should be refined in least-squares analysis.

4.6 Conclusion

Formulae have been given for the inclusion of absorption, multiple scattering, and extinction in Rietveld analysis. The two refinable parameters are D, which is the effective specimen size, and l, which is the size of the mosaic blocks. These should give refined values which are of the order of magnitude of the specimen size and mosaic block size respectively. The latter, particularly for brittle materials, is the grain or particle size shown by scanning electron microscopy.

If the refinements give values which are significantly different from these it is highly likely that the model is deficient.

References

Coppens, P. and Hamilton, W. C. (1970). *Acta Crystallogr.*, **A26**, 71–83.
Darwin, C. G. (1922). *Phil. Mag.*, **43**, 800–29.
Hamilton, W. C. (1957). *Acta Crystallogr.*, **10**, 629–34.
Hewat, A. W. (1979). *Acta Crystallogr.*, **A35**, 248.
Marshall, W. and Lovesey, S. W. (1971). *Theory of neutron scattering*. Clarendon Press, Oxford.
Read, W. T., jr. (1953). *Dislocations in crystals*. McGraw-Hill, New York.
Sabine, T. M. (1985). *Austral. J. Phys.*, **38**, 507–18.
Sabine, T. M. (1988). *Acta Crystallogr.*, **A44**, 368–73.
Sabine, T. M., von Dreele, R. B., and Jorgensen, J. E. (1988). *Acta Crystallogr.*, **A44**, 374–9.
Weinstock, R. (1944). *Phys. Rev.*, **65**, 1–20.

5

Data collection strategies: fitting the experiment to the need

Roderick J. Hill

5.1 Introduction

The abundance and complexity of the structural and physical information that can be extracted from powder diffraction patterns is approaching that obtainable from single crystal data. It is advisable, therefore, that careful consideration be given to the selection of the conditions under which powder data is collected and to the effect of these choices on the capabilities and results of the analysis.

The fundamental measured quantities (observations) for crystal structure refinement using powder diffraction data are the intensities of the Bragg peaks; the intensities collected at each step in the pattern serve only as multiple, variably-weighted, estimates of these values. The precision of the peak intensity measurement can be improved by increasing the number of counts accumulated at each step and/or the number of steps measured across the peak, but this is useful only up to the point where counting variance becomes negligible in relation to other sources of error; thereafter, time is wasted.

Furthermore, unlike single crystal diffraction, the projection of the three-dimensional reciprocal lattice onto the single dimension of a powder pattern often requires a debilitating compromise to be effected between the quality (dependent on resolution) and the quantity (dependent on wavelength and angular range) of the observations (peak intensities). The desire to increase the observations-to-parameters ratio must, therefore, be offset against the ability to discriminate accurately between the overlapping observations.

In this chapter, the criteria and means are presented by which sensible decisions can be made about the strategy for X-ray and neutron (particularly constant-wavelength) data collection, including the choice of instrument geometry, radiation and wavelength, step-scan range, step interval, and step intensity. The influence of these parameters on the observations-to-parameters ratio, on the most effective utilization of available time, and on the accuracy and precision of the derived crystal structural parameters is also discussed.

Although most of the comments are directed towards the optimization of the experiment for Rietveld analysis, they are also relevant to the first stage of pattern-decomposition studies wherein the integrated intensities are extracted without reference to a crystal structural model. Consideration is also given to the differences between X-ray and neutron data analysis, and to the effect of multiple phases in the sample.

5.2   Choice of diffraction instrument

The choice about which instrument to use for a particular experiment is often dictated more by what is conveniently available within a reasonable time-frame than by what would be best suited to the problem under investigation. For the refinement of moderately simple structures, a well-maintained, sealed-tube X-ray instrument capable of collecting step-scan data of moderate resolution (viz., about $0.08°\ 2\theta$) will usually suffice.

However, if the material under study has low symmetry, large numbers of variable atomic coordinates (i.e. more than about 50), heavy absorption, contaminant phases, a heavy/light atom mixture, split peaks, super-symmetry, or other very subtle structural details, then the choice of instrument (and other data-collection parameters; see below) may be critical in determining whether the outcome of the experiment is successful or not. Indeed, it may be immediately obvious that the problem is not solvable unless a specialist instrument operating from a synchrotron source (for maximum resolution) or neutron source (for refinement of light-atom parameters) is used. Even then, the detail of interest may be so subtle that multiple data sets may have to be collected on different instruments, or with different wavelengths, and the structure refined using all data sets simultaneously (Chapters 12 and 13 of this volume; Maichle *et al.* 1988; Williams *et al.* 1988).

Figure 5.1 shows the variation in instrument-only contributions to the peak full-width-at-half-maximum (FWHM) as a function of diffraction angle (i.e. the resolution function) for a number of modern neutron and X-ray diffractometers. As a general rule, the resolution of these instruments increases in the order neutron, conventional X-ray, synchrotron X-ray.

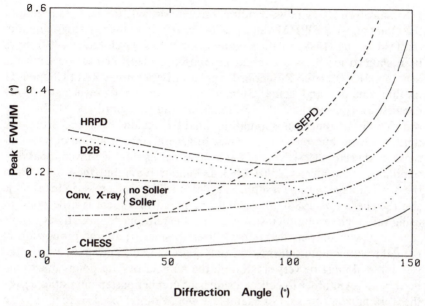

Fig. 5.1 Variation in the instrumental-only peak FWHM as a function of diffraction angle (2θ) for several neutron and X-ray diffractometers. The curves represent the following instruments: long — — — HRPD (CW neutron) at Lucas Heights; · · · D2B (CW neutron) at the ILL; short − − − SEPD (TOF neutron) at the ANL; ·−·−· conventional (X-ray) Bragg–Brentano with (lower) and without (upper) diffracted-beam Soller slits; —— synchrotron (X-ray) Bragg–Brentano at CHESS. Modified from Hill and Madsen (1987).

A partial exception to this is the neutron constant-wavelength (CW) machine D2B at the Institute Laue–Langevin (ILL), Grenoble, which has superior resolution in the high angle region, where the effect of a 135° monochromator take-off angle comes into play.

Neutron time-of-flight (TOF) instruments are not directly comparable to the CW machines since their resolution is usually stated in terms of a (constant for a particular detector-bank) value of $\Delta d/d$. The equivalent 2θ resolution of a machine with a $\Delta d/d$ value of 1.5×10^{-3} is plotted in Fig. 5.1 as the curve labelled SEPD (for the Special Environment Powder Diffractometer at the Argonne National Laboratory, ANL); it shows a resolution which is of the same order as X-ray instruments at low angles, but inferior to most CW neutron machines at high angles. (But see, also, Chapter 11.)

The figure also demonstrates the significant advantage that can be achieved in Bragg–Brentano diffractometers by the use of a set of diffracted-beam Soller slits limiting axial divergence, in this case, to the quite conservative value of 5°. Further improvements in resolution can be obtained with an incident-beam monochromator and/or Soller slits, or the use of very narrow receiving slits (Fawcett *et al.* 1988; Louër and Langford 1988).

Synchrotron X-ray powder diffractometers provide the best resolution at the moment, with FWHM values as low as 0.017° at the focusing minimum (Attfield *et al.* 1988). Values approaching this level (i.e. 0.035°) have been claimed for the most recent generation of TOF neutron instruments (e.g. the HRPD at the Rutherford Appleton Laboratory, RAL; Chapter 11 in this volume) and some Guinier and other single-wavelength X-ray geometries (Fawcett *et al.* 1988; Louër and Langford 1988), but the broadening becomes substantial in angular regions removed from the optimum focus and/or at high angles. In general, the resolution of position-sensitive detectors (PSD's) is of the same order as that of conventional (e.g. scintillation) counters (Wölfel 1983; Lehmann 1987), but they suffer from the disadvantage that the resolution is degraded by an increase in the sample size (Schäfer *et al.* 1984). It should be noted that there is little point in using very high resolution instruments for the refinement of relatively simple structures with only moderate peak overlap since (i) a large proportion of the data will contain information only about the background, and (ii) the step interval must be very small (and the number of data points must then be very large) in order to collect a sufficient number of step intensities across the (very narrow) peaks. For example, a step width of 0.004° is required (see below) for peaks with widths of 0.02°, and this leads to a total of 15000 observations for a scan range of only 60° 2θ. Of course, the diffractometer can be programmed to skip the inter-peak regions in a simple powder pattern (thereby saving data-collection and analysis time), but the fact remains that the high resolution of the instrument is neither necessary, nor utilized, in this case.

If the crystal structure is unknown, then single-wavelength data and the best possible resolution are essential:

(1) to detect minor impurity phases;

(2) to identify and quantify peak splitting and specimen broadening effects;

(3) to ensure the success of automatic indexing routines; and

(4) to ensure that a sufficient number of accurately-resolved reflections (structure factors) can be extracted for Patterson or direct methods of structure solution (Rudolf and Clearfield 1985; Lehmann *et al.* 1987; Attfield 1988; McCusker 1988). Attfield *et al.* (1988) have suggested that about 10 reasonable E values per atom in the asymmetric unit are required for a successful outcome of direct methods procedures.

For crystal structure refinement, the results obtained with CW neutron data tend to be more nearly accurate than those obtained with X-rays, for several reasons:

(1) the generally lower absorption coefficients for neutrons means that larger samples can be used (i.e. more crystallites are sampled), giving a better 'powder average' during data collection;

(2) the peak shape functions are generally simpler in (CW) neutron diffrac- tometers;

(3) neutron scattering lengths have essentially no angular dependence in the range of interest, so that the scattering power of the atoms does not diminish as rapidly at high 2θ angles; this means that meaningful data can be collected to much higher angles and the thermal parameters can be estimated much better.

In TOF neutron data, problems associated with the wavelength dependence of extinction and absorption effects, and with the accurate definition of background and peak shape have, in the past, rendered the thermal parameters less accurate, but recent developments (Chapter 11 in this volume) have largely solved these difficulties.

Ideally, the choice of instrument should be dictated by the purpose of the experiment, rather than availability, time and/or cost. Using a summary of the above arguments, the following rough criteria can be applied to make the selection:

Object/problem	Action
Structure solution	Use single-wavelength X-rays (i.e. a synchrotron source or a Guinier camera with an incident-beam monochromator), since resolution is of greatest importance.
Structure refinement (moderate complexity)	Use neutrons, since the atoms then have a more equal, and less 2θ-dependent, contribution to the overall scattering.
Structure refinement (very complex)	Use single-wavelength X-rays (as above) or latest-generation neutron TOF data, since resolution is generally superior.
Thermal parameters	Use neutrons since the scattering power of the atoms does not then decrease with $\sin\theta/\lambda$, as for X-rays.
Very subtle structural detail	Collect multiple data sets with X-rays and/or neutrons, and use simultaneous refinement methods.

Small sample Use X-rays (for which the scattering power is generally large) in association with Debye–Scherrer, thin-film, or Guinier geometry, for which sample volumes are minimal.

High X-ray absorption Use neutrons, or X-rays in conjunction with reflection (i.e. Bragg–Brentano) geometry.

Preferred orientation Use neutrons; for X-rays, avoid Bragg–Brentano geometry, and utilize Guinier or Debye–Scherrer geometry.

5.3 Choice of wavelength/resolution

Although Rietveld and pattern-fitting methods (Chapter 14) have been developed specifically to deal with the problem of peak overlap it is, nevertheless, desirable to collect a pattern with the best possible resolution. Unfortunately, this desire is in competition with the need to have a sufficiently large number of observations (i.e. Bragg intensities) for the task at hand. As indicated above, a useful rule-of-thumb for structure solution is that the ratio of resolved and observed reflections to the number of crystallographically independent atoms should be at least 10 (Attfield *et al.* 1988). For structure refinement, the parameter accuracy and precision may be acceptable only if the observations are in the majority by at least a factor of 5.

The number of peaks in any powder diffraction pattern is dictated solely by the size and symmetry of the unit cell and the wavelength of the radiation used. Thus, the number of reflections, N, up to the diffraction angle θ is given by

$$N \approx 32\pi/3 \; V/\lambda^3 \; \sin^3 \theta/Q \tag{5.1}$$

where V is the unit cell volume, λ is the wavelength, and Q is the product of the average multiplicity of the reflections and the number of lattice points per unit cell. The density of peaks per degree at 2θ is then given by (Christensen *et al.* 1985):

$$D \approx 4\pi^2/45 \; V/\lambda^3 \; \sin^2 \theta \cos \theta/Q. \tag{5.2}$$

For a triclinic material with unit cell volume of 1000 Å, the total number of reflections accessible to 180° 2θ using a wavelength of 1 Å is a staggering 17 000, and the maximum density of reflections is about 170 per degree at a 2θ value of 110° (Fig. 5.2). Of course, the total number of reflections accessed (and their density in 2θ) can be reduced if either the wavelength or the symmetry is increased. Figure 5.2 shows the 8-fold reduction in reflection number and density that follows in the specific case of a doubling of the

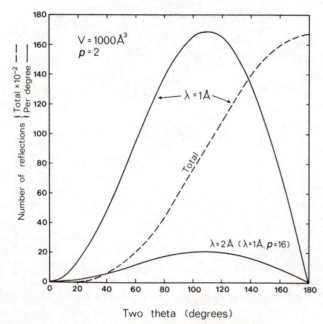

Fig. 5.2 Variation in the total number (dashed line) and density (continuous lines) of Bragg reflections in the powder pattern of a material with a unit cell volume (V) of 1000 Å³, as a function of diffraction angle (2θ) for two combinations of wavelength (λ) and reflection multiplicity (p).

wavelength or an increase in the symmetry from triclinic to tetragonal. Figure 5.3 gives the general relationship between the total number of reflections and the wavelength.

The appropriate choice of wavelength is a compromise between keeping the reflection overlaps (density) within manageable proportions by using a longer wavelength, and maintaining a reasonable observations-to-parameters ratio by using a shorter wavelength. Some of the options available for a sample of orthorhombic $PbSO_4$, with unit cell volume 318 Å³ and 16 crystal structural and 14 profile parameters to be determined, are summarized in Table 5.1.

In this case, a data-collection wavelength of 1.5 Å is more than adequate since the observations-to-parameters ratio is 12 and the reflection intensities are likely to be well-determined with the low reflection density of 4 per degree.

Indeed, if the available experiment time is short, a wavelength of 2.0 Å is probably satisfactory. However, if the upper limit of the scan is substantially smaller than 150°, then a shorter wavelength should be used to make up for the decrease in the number of accessed reflections.

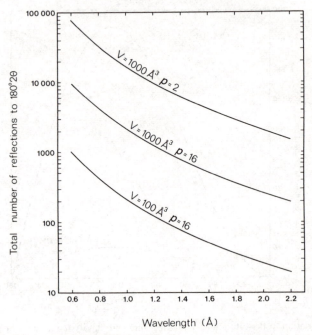

Fig. 5.3 Variation in the total number of reflections (to 180° 2θ) in the powder patterns of materials with various combinations of unit cell volume (V) and reflection multiplicity (p), as a function of wavelength.

Table 5.1 Effect of wavelength choice on the total number of reflections, the reflections to parameters ratio, and the reflection density for $PbSO_4$

λ (Å)	No. Reflns to 150° 2θ	Refln:Param ratio	Refln/° 2θ at 110° 2θ
1.0	1202	40.1	13.4
1.5	356	11.9	4.0
2.0	150	5.0	1.7

Another point, rarely considered, is that if the peak resolution (including shape definition) is not severely compromised by a decrease in wavelength, then data with the same observations-to-parameters ratio can be collected with the same step interval over a much shorter angular range. As a result, the total experiment time can be shortened at little or no cost to accuracy or precision.

The maximum peak density that can be tolerated will be a function of the diffraction peak widths, which are themselves dependent on the intrinsic resolution of the instrument used and the degree of crystallite-size and/or microstrain broadening due to the sample. There is, for example, not much point in having an instrumental resolution function that provides the narrowest peak widths at low angles, when the maximum peak density necessarily occurs at around 110°. For this reason, monochromators servicing the D2B and D1A diffractometers at the ILL (and other locations) have been specifically designed with high take-off angles that produce minimum peak widths at 2θ angles of 120° and 135°, respectively. This design feature has yet to be incorporated routinely in X-ray instruments since the peak intensities fall off rapidly with diffraction angle and so high-angle data are rarely collected.

The effect of severe crystallite-size broadening on the appearance of neutron diffraction patterns obtained from $PbSO_4$ is illustrated in Fig. 5.4. The simulated pattern in the lower part is identical to the observed pattern at the top, except for the simulated size broadening corresponding to the presence of 150 Å (rather than 1200 Å, as observed; Hill *et al.* 1984) crystallites in the sample. The resolution and quality of the lower pattern is degraded (especially at high angles), not only because the minimum FWHM is larger by a factor of two, but also because the step intensities are smaller by a similar amount.

A small gain in resolution may be achieved (if size-broadening does not dominate the total width of the peaks) by increasing the wavelength from, say, 1.377 to 1.893 Å (the appropriate cut-off for 1.893 Å is shown in Fig. 5.4) and thereby spreading a reduced number of reflections over the same accessible angular range. Clearly, in cases of line broadening as severe as that illustrated in Fig. 5.4, the high resolution of, say, a diffractometer operating from a synchrotron source cannot be exploited, and the use of such instruments is hard to justify.

For conventional X-ray data, the presence of the K_α doublet and its satellites imposes an additional limit on the resolution of the pattern. The use of single-wavelength data is to be highly preferred since it not only provides a halving of the reflection density at no cost to the observations-to-parameters ratio, but it eliminates systematic errors associated with the analytical removal of the α_2 component of the K_α doublet (Louër and Langford 1988).

All of the above factors are the reasons why, in the case of Rietveld analysis of very complex structures like the pentasil zeolites, it has often been necessary to complement the Bragg intensities with other pseudo-observations in the form of constraints applied to bond distances and angles, or to specific functional groups (Chapter 10 of this volume; Immirzi 1980; Pawley 1980; Baerlocher 1984; Rudolf and Clearfield 1985). As the complexity of the

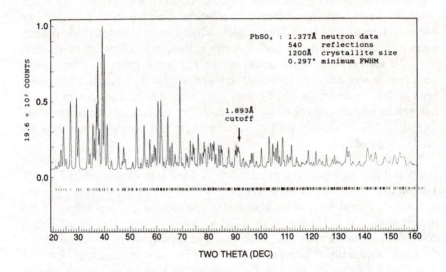

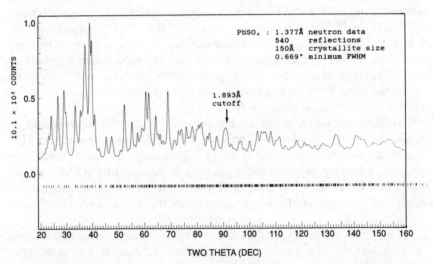

Fig. 5.4 Observed (upper) and calculated (lower) neutron powder diffraction patterns for anglesite, $PbSO_4$, corresponding to the presence of crystallites of mean size 1200 and 150 Å, respectively. The positions of all reflections accessible with a wavelength of 1.377 Å are indicated by tick marks below the patterns; the arrow shows the upper limit of reflections observable with a wavelength of 1.893 Å. From Hill and Madsen (1987).

structures coming under scrutiny continues to increase, the implementation of such constraints, together with the simultaneous analysis of separate data sets (Chapter 12 of this volume; Maichle *et al.* 1988; Williams *et al.* 1988), will find increasing importance.

5.4 Pattern analysis: basic requirements

If the ultimate aim of the analysis of a step-scan powder diffraction pattern is the determination and/or refinement of crystal structural parameters, then five fundamental requirements must be satisfied:

(1) the step intensities must be collected at known increments in 2θ or equivalent time-of-flight or *d*-spacing (Chapter 11);

(2) the centroids of the peaks must be reproduced by models involving the unit cell dimensions, 2θ-zero position, and instrumental aberrations (or the peak positions must themselves be variables);

(3) the shapes, widths, and asymmetry of the peaks must be adequately parameterized with reference to models for crystallite size and microstrain broadening, stacking faults, dislocations, etc. (Chapters 7, 8, and 11);

(4) the background must be suitably parameterized, or independently measured (Chapter 6);

(5) the peak intensities must represent the projection of the intensity-weighted three-dimensional reciprocal lattice onto the 2θ scale, modified perhaps by parameters accounting for preferred orientation and extinction.

The extent to which these requirements are realized is largely within the control of the experimenter if proper consideration is given to the interplay between instrument, sample, and model. They are here discussed in turn.

5.4.1 *Step-scan increments*

Knowledge of the 2θ step increment is generally not a problem with well-maintained conventional X-ray and neutron diffractometers, and film scanners. However, the peak widths obtained with some instruments on synchrotron X-ray sources are of the order of $0.02°$ 2θ; the stepping motors must then be capable of reproducibility at the level of $0.001°$, or less, and this cannot always be guaranteed.

For example, Lehmann *et al.* (1987) found an average positioning error of $0.006°$ in their synchrotron data, and attributed this both to mechanical error in the 2θ shaft and to instabilities in the linear position-sensitive detector. The resultant error in the intensity at half-maximum of these very narrow peaks was estimated at 10 per cent and led them to conclude that the local standard deviation of the step intensity can be much larger than the value obtained from quantum counting statistics. As pointed out by Prince (1985), a similar problem arises in multi-detector data where the interval between detectors is not a precise multiple of the step size. This issue is discussed further below.

5.4.2 *Peak positions, widths, and shapes*

The requirements (2) and (3) above are almost always violated to some degree or other in routine step-scan work, not only as a result of problems with instrumental alignment, but also due to unavoidable properties of the sample and the optics of the diffraction process (Klug and Alexander 1974). The nature of some of these aberrations, and ways in which their effects on the analysis of the pattern can be minimized, are summarized below for the common case of para-focusing geometry diffractometers on conventional X-ray sources with rectangular focal spots:

(1) X-ray source: view laterally with a take-off angle less than about 3° for highest resolution, and use an incident-beam monochromator to eliminate the K_{α_2} component (and resultant peak-stripping problems);

(2) $2\theta/\omega$ mis-set: causes peak broadening (maximum at low 2θ), and is eliminated by good alignment (ω should be $= \theta$);

(3) specimen flatness: reduce the X-ray beam footprint length on the sample to reduce peak shifts and asymmetric broadening (minor except at low 2θ);

(4) specimen displacement: causes peak shifts towards higher or lower 2θ (maximum at low 2θ) and is eliminated by good alignment;

(5) specimen transparency: low absorption causes asymmetric broadening and peak shifts (maximum at 90° 2θ) and is reduced with a thin sample (at cost to magnitude and angular dependence of intensity) or use of a longer wavelength (at cost to number of reflections);

(6) knife-edge slits: increase divergence or receiving slit widths to increase intensity, but reduce receiving slit width to increase resolution; decrease receiving slit length (i.e. decrease intersection with Debye–Scherrer cone) to reduce asymmetry at low and high 2θ;

(7) Soller/parallel slits: decrease axial/vertical divergence to minimize peak shifts and asymmetric broadening (maximum at low and high 2θ).

If any of the above aberrations have not been eliminated from the experimental set-up, or correction terms have not been included in the calculated models for peak shape and position, then severe problems may be encountered in the analysis of the pattern. Much effort has been expended in achieving better models for *peak shape* in CW and TOF diffractometry (Wertheim *et al.* 1974; Hall *et al.* 1977; Albinati and Willis 1982; Howard 1982; von Dreele *et al.* 1982; Young and Wiles 1982; Baerlocher 1984; David and Matthewman 1984; Greaves 1985; Hill and Howard 1985; Will *et al.* 1987; Chapters 7, 8, 9, and 11 in this volume). However, it is only rarely that the basic importance of correct modelling of the *peak positions* is acknowledged and given adequate priority (Thompson and Wood 1983; Retief *et al.* 1985).

The resolution of many new instruments, especially those using synchrotron radiation sources (Hastings *et al.* 1984; Parrish 1986; Thompson *et al.* 1987; Will *et al.* 1987), and conventional designs fitted with incident-beam monochromators (Fawcett *et al.* 1988; Louër and Langford 1988), is in the range 0.02–0.07° 2θ. Thus, the minimum FWHM values are of a similar magnitude to the shifts in peak positions produced by some of the more severe aberrations described in (1)–(7) above. This highlights the importance of the peak profile and position models used in pattern analysis. For this reason, new instrumental designs which reduce or eliminate peak-displacement errors by the use of parallel-beam (synchrotron) X-rays and an analyser crystal in place of the usual receiving slit (Thompson *et al.* 1987; Attfield *et al.* 1988) have much to recommend them. However, care must be exercised even with these instruments; Thompson *et al.* (1987) observed discrepancies of up to 0.05° between the observed and calculated peak positions, which they attributed to drifts and discontinuities in wavelength during decay of the stored electron current in the ring.

Of course, the sample itself can produce different peak widths and shapes for different classes of *hkl* indices in the diffraction pattern, due to anisotropic crystallite size and microstrain effects. In addition, stacking faults and incommensurate unit cell dimensions can produce significant non-uniform peak shifts. The model used in the analysis must then include parameters that account for these effects (e.g. Chapter 11; Greaves 1985), or the study of the full diffraction pattern is best abandoned in favour of measurement of the peak intensities by individual-peak profile analysis or planimetry.

5.4.3 *Background definition*

The definition of background can be a major problem in pattern analysis, especially when the range of diffraction angles does not extend beyond the region of maximum peak density near 110° 2θ. This is often the case for X-ray data, where the decay in the scattering factor magnitudes with $\sin \theta / \lambda$ sometimes makes the collection of data at high angles relatively fruitless. Unfortunately, it is in this high-angle region of the pattern that the peaks are broader and usually more Lorentzian in character (i.e. with longer tails), thereby making their distinction from the background even more difficult.

The situation is of less concern for CW neutron diffraction data since the atomic scattering lengths are effectively θ-independent and reasonably intense peaks, on a relatively low background, can be collected out to the mechanical limits of the instrument (generally 165° 2θ). At this limit, the peak density has declined somewhat and the intervals between the peaks may be sufficient to allow a reasonable estimation of the background level. On the other hand, TOF neutron data has a very intense background in the low d-spacing region, with a complex form corresponding to the spectrum of the incident flux (Von Dreele *et al.* 1982; Chapter 11 of this volume).

Another source of difficulty involves the definition of the range of influence of the peaks in a pattern. For a Gaussian peak, 99 per cent of the area is within 3 FWHM of the centroid, but for a Lorentzian peak the same proportion of the peak area is included only if the range of influence is extended to 63 FWHM (Toraya 1985). Even for a peak of 'intermediate' Lorentzian character, some 9 FWHM are required for 99 per cent of the peak area to be considered. Since each peak contributes intensity over a large range of diffraction angles on either side of its centroid, the usual situation is that large numbers of peaks contribute intensity to any given step in the pattern. The increase in computing time that necessarily results from the need to iterate over this large number of peaks is not tolerated by some users, with the result that the truncated parts of the peaks are assimilated into the background. This leads to corresponding errors in the thermal parameters and, for low-symmetry materials, the atomic coordinates (Toraya 1985).

The proper definition of the background can also be difficult if the sample contains a significant proportion of amorphous, or very poorly crystalline, material. Successful attempts to model the irregular backgrounds arising from amorphous material have involved the use of fifth-degree polynomials in d or θ (Baur and Fischer 1986), and Fourier filtering techniques (Richardson *et al.* 1988), the latter of which can be used to extract a real-space function corresponding to the atomic radial distribution function of the amorphous phase. See Chapter 6 for a more extensive consideration of background modelling due to the Fourier filtering technique.

5.4.4   Peak intensities

For crystal structure solution and refinement the measured peak intensities must represent a known mapping of the intensity distributions of all *hkl* reflections onto the 2θ (or 'd') dimension of the powder pattern.

In the simplest case, this mapping is based on the assumptions (i) that the sample is composed of randomly oriented crystallites (i.e. no preferred orientation), and (ii) that these crystallites are sufficiently numerous to present essentially all possible orientations to the incident beam (i.e. a 'powder average'). Many methods have been proposed to ensure that these requirements are met, including the use of randomizing sample movements, homogenization with amorphous diluents, and crystal size reduction (Klug and Alexander 1974).

The degree to which a randomly oriented distribution can be achieved varies with the instrument used. Guinier cameras, for example, can provide data essentially free of preferred orientation effects, whereas para-focusing Bragg–Brentano instruments are notorious for the presence of sometimes severe non-random distributions. The problem of obtaining a powder average is especially troublesome for instruments on synchrotron sources since the parallel beam geometry implies that a very restricted number of crystallites have the correct orientation to reflect. In the latter case, reduction in crystallite size, together with sample rotation and rocking motions is advisable. Details of the means by which software corrections for preferred orientation can be made during pattern analysis are provided in Appendix 5.A to this chapter.

Due to the small crystallite sizes used for powder diffraction, extinction does not generally influence the relative peak intensities to a major extent (Spackman *et al.* 1987). However, materials with high symmetry can display significant extinction, even in polycrystalline form and an accurate model for its correction has been provided (Sabine 1988; Chapter 4 and Appendix 5.A to this chapter).

5.5 Choice of step width and step intensity

Having selected the instrument, wavelength, and scan range to be used, one may next decide on the step interval and step counting time (i.e. the overall intensity of the pattern) to be used during data collection itself. For the X-ray pattern of corundum, α-Al_2O_3, displayed in Fig. 5.5, the choices were: a step size of $0.01°$, a step counting time of 5 s, and a scan range of 24–$129°$ 2θ. This combination provides a total of 50 Bragg peaks, 10 480 data points (steps), a maximum step intensity of about 60 000 counts for a Bragg–Brentano diffractometer using a Cu tube operated at a power rating of 1.3 kW, and a total collection time of 14.5 h.

Rietveld analysis results have, however, been published in which the step widths range over two orders of magnitude, from 0.002 to $0.3°$ 2θ (Khattak and Cox 1977; Bacon and Lisher 1980), and in which the maximum step intensities range over nearly three orders of magnitude, from several hundred to many tens of thousands of counts (Immirzi 1980; Hill 1982). Despite this wide range of experimental conditions, very little, if any, discussion has been presented about the reason for their choice, or about their possible effect on the results of the refinement.

For example, Table 5.2 provides a small selection of reasonable combina-

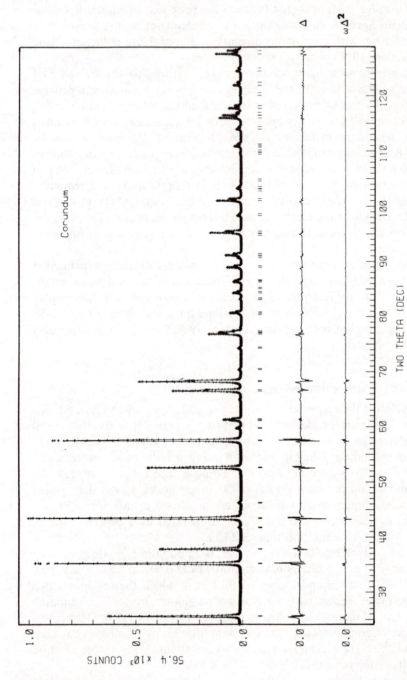

Fig. 5.5 Observed, calculated, and difference Cu K_α X-ray powder diffraction patterns for corundum, α-Al$_2$O$_3$. The observed data, collected at a step interval of 0.01° 2θ and a step counting time of 5 s, are indicated by plus signs, and the calculated pattern is the continuous line overlaying them. The tick marks below the pattern represent the positions of all possible Bragg reflections. The two lower curves show the value of Δ and sign$(\Delta)w\Delta^2$, respectively, at each step, where Δ is the difference between the observed and calculated intensity, and w is the weight applied during the least-squares refinement. From Hill and Madsen (1987).

Table 5.2 Effect of several choices of step width and step counting time on the maximum step intensity and total counting time of the corundum Cu K_α powder pattern

Step width (° 2θ)	No. of steps	Step time (s)	Max. step int. (counts at 1.3 kW)	Counting time
0.01	10 480	50.0	600 000	145.6 h
0.05	2096	5.0	60 000	2.9 h
0.2	524	0.5	6 000	4.4 min
0.1	1048	0.05	600	52 s

tions of step width and counting time that could have been used to collect the corundum pattern in Fig. 5.5. At one extreme, the set of 50 Bragg reflections could be collected with a step width of 0.01° and a step-counting time of 50 s, to yield 10 480 steps, a maximum step intensity of 600 000 counts, and a total experiment time of 146 h. At the other extreme, a step interval of 0.1° and a step counting time of 0.05 s produces 1048 steps with maximum intensity 600 counts, but now the pattern is collected in only 52 s of actual counting time (i.e. excluding the overheads required to move the detector between steps).

In the case of TOF neutron data, the situation is a little more complicated since the step intervals in the pattern relate to TOF rather than 2θ. When the TOF steps for the 150° data bank attached to the SEPD at the ANL, operated in high resolution mode (ΔTOF = 5 μs), are referenced to the 2θ scale for a CW instrument using 1.5 Å neutrons (Table 5.3), the equivalent 2θ intervals range from 0.013° at 40° 2θ, to 0.556° at 160° 2θ. These large TOF angle-equivalent step intervals at small d values (i.e. at the equivalent of high diffraction angles) are more than 10 times those in the CW case and are consistent with the relative angular resolution functions of these instruments given in Fig. 5.1.

Irrespective of the specific choice of data collection instrument, the question is: *What are the effects of changing the number of steps and the step intensity on the accuracy and precision of the refined structural parameters?*

Before the choice of step width and intensity is made, it should be remembered that the fundamental measured quantities (observations) for crystal structure solution and refinement are the intensities of the Bragg peaks; the intensities collected at each step in the pattern serve only as multiple, variable-weighted, estimates of these values. The precision of the peak intensity measurement can be improved by increasing the step intensities and/or the number of steps across the peak (i.e. decreasing the step interval), but this is useful (i.e. translates into a corresponding improvement in

Table 5.3 Comparison of step widths for a constant wavelength (CW) diffractometer, with equivalent values for a time-of-flight (TOF) instrument

$\lambda = 1.5$ Å		CW $\Delta 2\theta = 0.05°$		TOF $\Delta TOF = 5$ µs	
2θ (°)	d (Å)	$\Delta 2\theta$	Δd	$\Delta 2\theta$	Δd
40	2.19	0.05	0.0026	0.013	0.00066
160	0.76	0.05	0.000058	0.556	0.00066

parameter accuracy) only up to the point where counting variance becomes negligible in relation to other sources of error. *Beyond this point, no additional crystal structural information is obtained by further increasing the number of the steps or the number of counts collected per step.*

If the pattern analysis involves the method of structure-independent pattern-decomposition (PD), the step-scan data are discarded once the integrated intensities have been extracted (Pawley 1980; Cooper *et al.* 1981; Jansen *et al.*1988; Chapter 14 of this volume). In this case, if the intensities have been over-determined, then a waste of time is the only consequence. However, if the analysis is of the Rietveld kind, using a weighting scheme based entirely on quantum counting statistics, then other repercussions can arise.

Firstly, it is relatively rare that counting variance is the largest source of error in a step-scanned diffraction pattern since:

(1) there is almost always a problem with the peak shape and peak intensity (crystal structural, preferred orientation, etc.) models;

(2) the local error in the step intensity may be significantly larger than that provided by counting variance, due to counter mis-positioning (Lehmann *et al.* 1987) and/or to a non-integral number of steps between the detectors of a multi-detector array (Prince 1985);

(3) the variances of the step intensities rarely include the effect of the monitor and absorption factor corrections.

Furthermore, no account is taken of the fact that (when the background is not ignored) the relative weights assigned to the tails and tops of peaks change dramatically with changes in the signal-to-noise ratio.

While the detail of the weighting scheme is likely to have a negligible effect on the accuracy of the derived structural parameters (Chapter 3 of this volume), it is desirable, if for no other reason than felicity, that the weights be as close as possible to a realistic estimate of the reciprocal of the variance of the observations.

Secondly, if counting statistics are indeed negligible in relation to these other sources of error (i.e. if one or more aspects of the model are inadequate), the residuals across a given Bragg peak will be serially correlated (Hill and Flack 1987; Schwarzenbach et al. 1989).

One of the assumptions upon which the validity of the normal least squares method is based is that the observations are drawn at random from populations whose means are given by the model with the correct values assigned to all adjustable parameters, so that the error terms have zero mean. The presence of serial correlation in the residuals is an indication that this condition is not satisfied. The least-squares procedure then breaks down in two ways:

(1) the parameter estimates may be biased;

(2) statements of confidence based on t and F distributions are no longer valid.

Thirdly, a further complication arises in pattern analysis because, unlike the case of single-crystal data refinements, the number of observations in the step scan can be made arbitrarily large (independently of the number of Bragg reflections) simply by decreasing the step interval. The definition of the estimated standard deviation (e.s.d.) of a parameter requires, however, that it decreases as the square root of the increase in the number of observations (International Tables for X-ray Crystallography 1967). Thus, if the step interval is made small enough (*and the step-counting time remains constant*), the e.s.d. based on counting variance in pattern analysis can be made to approach zero.

At this point, it should be stressed that the e.s.d. is a measure solely of the *precision* of the parameter estimates. In the same way that good agreement between observations and model does not imply accuracy (Schwarzenbach et al. 1989), it should likewise never be construed that a small e.s.d. value is an indication of a more nearly accurate parameter.

Thus, the generally accepted and widely executed, but formally incorrect, practice of using the e.s.d. value to attach a certain level of significance (i.e. accuracy of probable error) to the numerical differences between estimates of the same parameter from independent measurements has little meaning. Indeed, if decreasing the step interval does not result in an improvement in the precision of the integrated intensity measurements, it is easy to see that any conclusions about the relative merits of different parameter estimates could be changed (in fact, reversed), at will, by the resultant inevitable decrease in the relative e.s.d. values.

In general, the breakdown in the validity of the weighting scheme and hence the derived e.s.d.'s will be indicated by an increase in the value of the goodness-of-fit index (Young *et al.* 1982) above its ideal value of unity (Scott 1983; Hill and Madsen 1984, 1986; Schwarzenbach *et al.* 1989).

When it is found that the data are serially correlated, one has the option of modifying the weighting scheme (Rollett 1982, 1988; Baharie and Pawley 1983; Cooper 1983; Prince 1985), adjusting the variances at the conclusion of the refinement by multiplying by the goodness-of-fit value (Pawley 1980; Scott 1983), or determining the *true e.s.d.'s* by replications of the experiment (Hill and Madsen 1984, 1986). None of these options is particularly desirable or advisable (Schwarzenbach *et al.* 1989). Alternatively, the data-collection conditions themselves can be adjusted to *retain the maximum information-content of the observations* while at the same time *ensuring that time is not wasted.*

With these latter two objects in mind, systematic studies have been undertaken to determine the effect of variations in step width and intensity on the accuracy and precision of the parameters derived in Rietveld analysis (Hill and Madsen 1984, 1986, 1987). In this work, X-ray powder diffraction data were collected on corundum over a wide range of step counting times and step intervals (Table 5.4).

Table 5.4 Summary of relevant crystal structure details for corundum, α-Al_2O_3, and of the ranges of step width and step counting-time utilized in comparative tests by Hill and Madsen (1984, 1986)

Corundum	Space group $R\bar{3}c$
Atomic coordinates:	Al at $(0, 0, z)$
	O at $(x, 0, 1/4)$
Unit cell:	$a = 4.76$ Å, $c = 12.99$ Å
	$V = 255.1$ Å^3
Number of reflections to 140° 2θ:	108 (for Cu K$_\alpha$)
Maximum reflection density:	1.3 deg^{-1}
Step width:	0.01–0.32° 2θ
Step counting time:	0.01–5 s
Step intensity:	120 to 58 000 counts

In assessing the relevance of the final recommendations of these studies, it should be remembered that they apply only in the context of defining the optimum data-collection conditions for refinement of the *crystal structure of a single-phase sample* displaying no superstructure or other higher-level effects. Furthermore, it is also assumed that the details of the line shapes are of no interest in their own right.

5.5.1 *Information content of the step-scan pattern*

5.5.1.1 *Structure refinement result* The results of crystal structure refinements undertaken on corundum are shown in Fig. 5.6. In the top part of the figure, the step width is fixed at 0.04°, while the step counting time varies over a factor of 500. In the bottom part, the counting time is fixed at 5 s per step, while the step width varies by a factor of 32. In both cases, the refined values of the oxygen atom x-coordinate do not depart significantly from the single-crystal value of Lewis *et al.* (1982), marked LSF.

The e.s.d. of the parameter determination (indicated by error bars) increases with decreasing step counting time and with increasing step width;

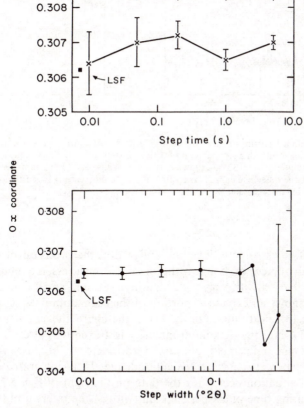

Fig. 5.6 Variation in the value of the x-coordinate of the oxygen atom in corundum determined from Cu K_α X-ray powder diffraction data collected over a range of step counting times (upper) and step widths (lower). The e.s.d.'s of the parameter values are indicated by error bars, and the single-crystal value of Lewis *et al.* (1982) is marked by a filled square and labelled LSF. From Hill and Madsen (1987).

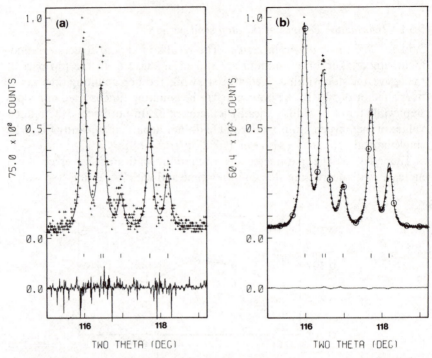

Fig. 5.7 Expanded portions of the corundum Cu K_α X-ray powder pattern near 117° 2θ (cf. Fig. 5.5), collected with a step interval of 0.01° and step counting times of (a) 0.05 s, and (b) 5 s. Data at intervals of 0.32° are shown in (b) as circled points. The patterns are as described in Fig. 5.5; the difference curve is the value of Δ_i, the difference between the observed and calculated intensity at each step i. From Hill and Madsen (1987).

at these extremes of counting time and width, the precision of the Bragg peak intensity determination is rather poor. Nevertheless, the refinements all proceed well, even with very wide step intervals and small counting times. Figure 5.7 shows an expanded portion of the corundum data set near 117°, collected at these extremes. On the left of the figure, the step counting time is only 0.05 s, and the maximum intensity in the entire pattern is only 600 counts, yet the refinement converged satisfactorily because a sufficient number of steps were collected to define adequately the integrated intensities.

A similar situation occurs for the data on the right of Fig. 5.7, obtained with a counting time of 5 s per step, but with a step interval of 0.32° (data points at intervals of 0.01° are shown for clarity only; those at a spacing of 0.32° are circled). Each of these widely-spaced step intensities is well determined by the 5 s counting time and thus it is not necessary to collect many steps across the peaks. Note that at this extreme (as for the case above), the peaks are poorly defined to the eye, yet the refinement proceeded quite satisfactorily.

5.5.1.2 *Bragg agreement index* The concept of saturation of the peak intensity measurement is illustrated by the plot of Bragg agreement index, R_B, as a function of step width and counting time (Fig. 5.8). This index is the one that most closely reflects the fit of the crystal-structural model to the observations (Young *et al.* 1982). For the data collected with a counting time of 5 s per step, the value of R_B remains virtually constant at step widths up to 0.16°. This indicates that no additional useful information is obtained about the integrated peak intensity value when the data are collected at step intervals narrower than 0.1°. Since the minimum peak width for this diffraction pattern is 0.17° 2θ, Fig. 5.8 suggests that a step interval of about half the FWHM value is sufficient to adequately define the peak intensities when the counting time is quite long (i.e. 5 s, representing a maximum step intensity of about 60 000 counts).

On the other hand, the R_B value for the data collected at 0.05 s per step falls continuously as the step interval is decreased. This shows that the integrated intensity measurement is not completely saturated at any of the step widths chosen and that further structural information (although not a

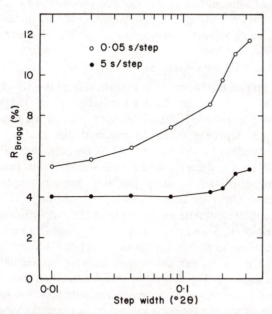

Fig. 5.8 Variation in the Rietveld Bragg agreement index, R_B, as a function of step width for refinements of the corundum structure using X-ray data collected with step counting times of 0.05 and 5 s. From Hill and Madsen (1986).

great deal) would be obtained if the step interval were less than 0.01°. In other words, *a high scan-point density is necessary only if step counting times are short.* Similar results have been reported by Will *et al.* (1988) in studies of quartz with the pattern decomposition technique, where only two points per FWHM were found to define the integrated intensities adequately for this purpose.

5.5.1.3 *Fourier transformation of the diffraction pattern*

It is not necessary to undertake a complete Rietveld analysis to see that the information content of step-scan data rapidly reaches a point of diminishing return. Cameron and Armstrong (1988) have examined the Fourier transform of the peaks in the so-called 'hand of quartz', using data collected at various step intervals between 0.01 and 0.1° 2θ (Fig. 5.9). As the step width is decreased, the diffractograms become more continuous and their transforms extend further into the Fourier domain. However, at a step width of 0.05° the transform has decayed to the baseline, indicating that the smaller sampling intervals serve only to provide information about the noise in the pattern.

The optimum sampling interval of 0.05° corresponds to the measurement of only 2–3 points per resolution element (i.e. the true minimum FWHM), well before the peak profiles have become smoothly outlined to the eye. For broader peaks, the resolution of the instrument could be lowered, or the step interval increased, without loss of information. The resultant (often substantial) savings in resolution and/or time can then be used to increase the signal-to-noise ratio, or to allow the instrument to be used for another problem.

5.5.2 *Estimated standard deviations*

A specific example of the variation in a particular e.s.d. with changes in step width has been provided by Rietveld refinements of the oxygen atom *x*-coordinate in corundum (Hill and Madsen 1986) using data collected at 5 and 0.05 s per step (Fig. 5.10). As expected, the e.s.d.'s for both step-counting times decline by the factor $\sqrt{32}$ as the step interval is decreased from 0.32 to 0.01°. The decrease in the e.s.d. values shows that the estimate of the oxygen coordinate is more *precisely* measured with the smaller sampling interval. However, does the change in step interval also provide a more nearly accurate estimate of the value of the *x*-coordinate?

The goodness-of-fit, S, and pattern indices R_{wp} and R_p (Young *et al.* 1982) are totally insensitive to changes in the step interval (Hill and Madsen 1986). Only the Bragg index R_B can distinguish between refinements at different step widths; as shown in Fig. 5.8, R_B at first declines and then 'bottoms out' as the step width decreases. It is reasonable (though not legitimate in a statistical sense, unless the model is known to be correct) to infer that, while the value of R_B is declining, the determination of the *x*-coordinate is becoming more nearly accurate. If this were not the case, then crystal-

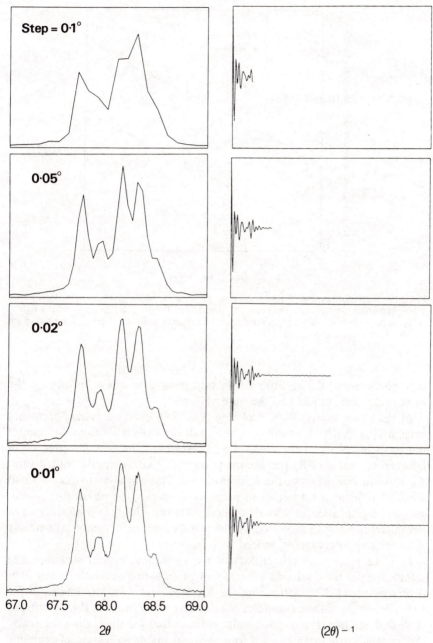

Fig. 5.9 Expanded portions (left column) of the quartz, SiO_2, Cu K_α X-ray powder diffraction pattern near 68° 2θ, collected with step widths between 0.1 and 0.01° 2θ (as indicated), and their corresponding Fourier transforms (right column). Adapted from Cameron and Armstrong (1988).

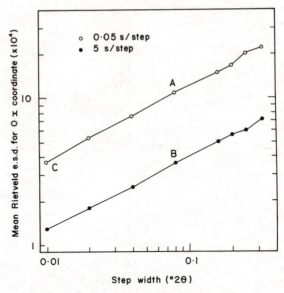

Fig. 5.10 Variation in the Rietveld analysis e.s.d., as a function of step width, for the oxygen-atom x-coordinate in corundum, determined from Cu K_α X-ray powder patterns collected with step counting times of 0.05 and 5 s. Points A, B, and C are as described in the text. From Hill and Madsen (1986).

lographers would derive little satisfaction from their quest to improve the fit between their model and the observations.

At the same time as R_B is declining (Fig. 5.8), the e.s.d. is also becoming smaller (Fig. 5.10), so in this case the e.s.d. is clearly a *de facto* estimate of the probable error in the x-coordinate. However, the tenuous connection between the value of R_B (i.e. accuracy) and the e.s.d. is strictly limited since R_B bottoms out, whereas the e.s.d. does not. Thus, in a comparison of two Rietveld refinements, there is no point in discussing the physical/crystallographic significance of the differences between the two estimates of a parameter in terms of their e.s.d.'s if the two Rietveld refinements have grossly different step intervals (or, indeed, step intensities).

In these systematic refinements of the corundum crystal structure, the significance of the e.s.d.'s as estimates of probable error has been independently determined by comparing the Rietveld e.s.d.'s with estimates of the (true) precision of the parameters obtained by replications of the experiment (Hill and Madsen 1986). The results indicate that for short counting times, the Rietveld e.s.d.'s can be used to give a reasonable indication of probable error over the entire range of step intervals up to about the minimum FWHM value, whereas for long step-counting times, the e.s.d.'s can be so used only for a narrow range of step widths at, or just below, the FWHM.

5.5.3 Durbin–Watson d-statistic

The results of refinements undertaken with data sets collected at different counting times and step widths can also be assessed with the so-called Durbin–Watson d-statistic (Durbin and Watson 1971; Hill and Flack 1987). This statistic provides a quantitative measure of the serial correlation between adjacent points in a diffraction pattern. It has a form akin to that of a conventional crystallographic R-factor, except that it is calculated from a summation over the differences between adjacent residuals, Δy_i:

$$d = \sum_{i=2}^{N} (\Delta y_i/\sigma_i - \Delta y_{i-1}/\sigma_{i-1})^2 \Big/ \sum_{i=1}^{N} (\Delta y_i/\sigma_i) \tag{5.3}$$

where $\Delta y_i = y_i - y_{ic}$, $\sigma^2 = y_i$, and i is one of N steps in the pattern. A formula for testing the values of d (Theil and Nagar 1961) shows that the variances and covariances of the parameter estimates can be considerably in error when the values of d deviate from a narrow band around 2 (Schwarzenbach et al. 1989).

Figure 5.11 provides an illustration of the change in d as a function of

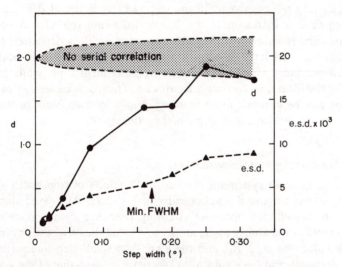

Fig. 5.11 Variation in the Durbin–Watson statistic, d, (circles; scale on the left) and the e.s.d. of the oxygen-atom x-coordinate (triangles; scale on the right) in corundum, as a function of step width, determined by Rietveld analysis of Cu K_α X-ray powder data collected with a step counting time of 5 s. The shaded area corresponds to the region into which the d value must fall if there is no serial correlation in the residuals at the 0.1% significance level. The arrow represents the minimum observed FWHM of the peaks in the pattern. From Hill and Madsen (1987).

step width for the 5 s corundum data refinements. The shaded area of the figure is the region into which d must fall if there is no serial correlation at the 0.1 per cent significance level. As the step interval widens, the intensities y_i are gradually rendered more independent by the fact that fewer points lie on each Bragg peak. However, it is not until the step width reaches a value of about $0.24°$ that the y_i values become truly independent. This indicates (Schwarzenbach *et al.* 1989) that the calculated e.s.d.'s approximate well the actual probable errors only for the refinements with data collected at step intervals at least as wide as $0.24°$, slightly larger than the peak FWHM. Note that, as is shown in Fig. 5.6, the accuracy of the crystal structural parameters has not been reduced in the process of increasing the step interval up to about $0.20°$. The extent to which the step interval can be widened without loss of information is, however, dependent on the inherent complexity of the pattern, as is explained elsewhere in this chapter, and the type of information sought.

Before proceeding further, it is worth examining another implication of the data in Fig. 5.10. Consider a data set collected under conditions corresponding to point A, that is, to a step interval of $0.08°$ and a counting time of 0.05 s; the e.s.d. of the oxygen atom x-coordinate is then about 0.001. This e.s.d. could be decreased by a factor of about three, say, to the level of the points B and C, in two quite different ways.

In following the path A to B, the step interval is left at $0.08°$, while the counting time is increased to 5 s, but in following the path A to C, the counting time remains at 0.05 s, while the step width is decreased to $0.01°$. Although the final level of parameter precision is the same in both cases, path AB produces a 100-fold increase in experiment time, while path AC increases the total time by only a factor of 8. Thus, *a given level of parameter precision can be achieved much more efficiently by a decrease in step width than it can by an increase in step counting time.*

5.5.4 *Summary and recommendations: step intensity and step width*

The results of these systematic studies of the effects of step width and step intensity are summarized schematically in Fig. 5.12. At a fixed step width, shown on the left, the optimum counting time is a compromise between (i) a decrease in the ability of the model to predict the observations (reflected in large values for R_{wp}, R_B, and the e.s.d.'s) at small step intensities (short counting times), and (ii) a substantial increase in the value of the goodness-of-fit statistic (indicating that the model is deficient and that the variances of the parameters are of dubious significance) and an unnecessarily long experiment time at large step intensities. Thus, in order to maximize the information-content of the observations and, at the same time, the efficiency of data collection, *the experiment should be designed so that between 5 and*

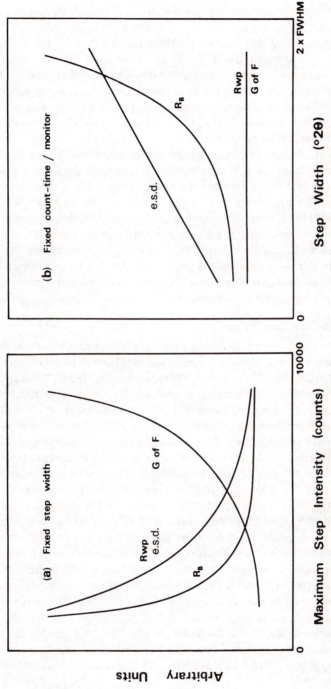

Fig. 5.12 Schematic representation of the variation in Rietveld agreement indices R_{wp} and R_B, goodness-of-fit index GofF, and parameter e.s.d., as a function of (a) maximum step intensity, and (b) step width. Only the extreme values have been marked on the abscissa scale since it is the intention to delineate trends in the indices, rather than specific detail. (From Hill and Madsen 1987.)

10 thousand counts are accumulated for the step intensities in the largest peaks in the pattern.

At a fixed counting time, shown on the right of Fig. 5.12, the optimum step width is a compromise between (i) a decrease in the usefulness of the e.s.d.'s as estimates of the probable error and wasted experiment time at small step intervals, and (ii) a decrease in the ability of the model to predict the observations (as measured by R_B) at large step widths. As a rule-of-thumb, *the optimum step width is between one-fifth and one-third of the minimum FWHM of the peaks.*

In the light of these results, it would be advantageous if a preliminary step-scan were made on the sample prior to the final data collection, to determine the maximum step intensity, minimum peak width, optimum wavelength, and maximum meaningful extent of the data in 2θ. This minimum FWHM for non-overlapping peaks should be determined at the angle corresponding to the maximum resolution of the particular diffractometer, and should take account of peak splitting, asymmetry and the like, if present. If data collection time is important, this can be reduced at little or no cost to precision, by choosing a short counting time and compensating for this with a narrower step interval.

5.6 Pattern-decomposition analysis

For *ab initio* structure solution it is necessary to extract the values of the integrated intensities by structure-independent, pattern-decomposition (PD) methods (Pawley 1980; Cooper *et al.* 1981; Jansen *et al.* 1988). Although PD and Rietveld analysis use similar procedures for fitting individual peak profiles, they are fundamentally different in the manner in which they extract the intensities from the full step-scan pattern. In PD the intensities of all but the exactly overlapping reflections are treated as separate refinable variables, whereas in Rietveld analysis the intensities are modelled in terms of the more limited number of variables provided by the crystal structure model (the methods are discussed in more detail in Chapter 14 of this volume).

There may also be a difference in the manner of fitting the peak positions; in the so-called Parrish-refinement method of PD, the 2θ location of every peak may be a separate variable rather than allowing them to be linked to the (maximum of six) unit cell parameters. Thus, the PD model generally has a larger number of parameters to be determined, and hence a larger number of degrees of freedom, than Rietveld refinement.

The implications of the greater degree of freedom (i.e. flexibility) in the PD method are that the step interval should be smaller, and that the step intensity (count time) should be larger in PD than they are for Rietveld analysis. In fact, unlike Rietveld analysis, the use of excessively long counting times or excessively narrow step intervals in PD has no consequences other than wasting time.

5.7 Neutron and X-ray differences

In recent years, the resolution and incident-beam flux of many neutron diffractometers has improved to the point where problems with the peak-shape model are becoming increasingly obvious. In these cases, as for data collected on X-ray instruments, quantum counting variance may no longer be the largest source of error in the Rietveld refinement. Thus, the same degree of care must be exercised in the choice of data collection parameters for neutrons as for X-rays.

In fact, the rules-of-thumb detailed above for choosing the step width and step intensity in X-ray data collection can be applied equally well to the case of CW neutron data. Thus, a step width of about 0.05° is generally sufficient for neutron diffractometers with a resolution (i.e. minimum FWHM) of around 0.2 to 0.3° 2θ, and monitor times should be set to provide maximum step intensities (summed over all counters contributing to the measurement) of the order of 5–10 thousand counts.

For instruments such as the D2B diffractometer at the ILL, Grenoble, the incident neutron flux is very high (by CW neutron standards) and each counter in the multi-detector array is only 2.5° away from its nearest neighbour. As a result, data from one counter generally provides sufficient intensity for the analysis, and the total collection time then rarely needs to be extended beyond 2 or 3 h, except for very weak scatterers, or very small sample volumes.

TOF neutron instruments also have a high incident flux, but the situation is a little more complicated since, as indicated earlier, the steps are in fixed units of TOF. The equivalent CW step width in 2θ varies (for $\lambda = 1.5\,\text{Å}$) from about 0.01 to 0.6° for a 5 μs TOF step (Table 5.3). This corresponds to the measurement of between 7 and 3, respectively, steps across the FWHM of the instrumental peak (Fig. 5.1), about the same as that in the CW neutron case.

5.8 Multiphase mixtures

Powder diffraction patterns often contain peaks from more than one crystalline phase. These extra peaks may arise from the sample containment, unreacted starting materials, decomposition or transformation products, materials added for dilution purposes, or non-separable phases in the original sample.

In the past, these 'impurity' peaks have commonly been treated as excluded regions during the pattern analysis; this practice is undesirable since it can lead to the loss of reflections of key interest, and to problems associated with peak truncation at the boundaries of the excluded regions. Thus, most Rietveld programs now allow for the presence of multiple phases by permitting the simultaneous refinement (or at least calculation of the diffraction patterns) of all phases in the sample (Worlton *et al.* 1976; Werner *et al.* 1979; Hill 1982; Howard *et al.* 1982; Bendall *et al.* 1983; Hill *et al.* 1984).

Hill and Howard (1987) showed that in binary mixtures of rutile, corundum, quartz, and silicon, the derived crystal structure parameters and unit cell dimensions are statistically independent of their abundance, at least down to the level of 10 wt.%. Indeed, the power of Rietveld analysis is such that Kisi *et al.* (1989) have determined the crystal structure of the ortho-rhombic form of zirconia using neutron data collected from a mixture of this phase with *four* other components. Note that the specific use of Rietveld analysis for the determination of phase abundance is described in Appendix 5.A.

The presence of multiple phases in the sample inevitably leads to an increase in the density of the diffraction peaks, and it may then be advisable to consider 'spreading out' the peaks by using radiation with a longer wavelength. In any event, the general rules-of-thumb for choosing step width and intensity proposed above should be relaxed somewhat in deference to the greater complexity of a multi-phase pattern. Just how much these rules should be relaxed has not been tested and will be strongly dependent on the particular combination of phases present.

It is also worth noting that the extent of the relaxation should be greater in the case of pattern decomposition studies than for Rietveld analysis. Consider, for example, the Cu K_α diffraction pattern from a 1:1 by weight mixture of cubic and tetragonal zirconia, shown in Fig. 5.13. The tetragonal phase is only slightly distorted from the cubic form, so that none of the cubic peaks are completely free-standing. Rietveld analysis can successfully deal with this mixture since it can use the information from free-standing tetragonal peaks to assist in the partitioning of the intensity in the groups where the tetragonal and cubic peaks overlap. Pattern-decomposition, on the other hand, cannot use this structure-based information to help in the fitting process, so all of the information must be extracted from the shapes and widths of the peak profiles alone. To assist this process, the step widths should be narrower, and the step intensities larger in PD than for a Rietveld analysis of the same mixture.

5.9   Acknowledgements

This material was prepared during tenure of a Ludwig Leichhardt Fellowship at the Mineralogisches Institut, Universität Würzburg, Germany; the author is most grateful to the Alexander von Humboldt-Stiftung (Germany) for their generous support. Most of the systematic studies of the effects of step width and step intensity were undertaken earlier in close association with Mr I. C. Madsen of the CSIRO Division of Mineral Products, whose past and present collaboration is greatly appreciated.

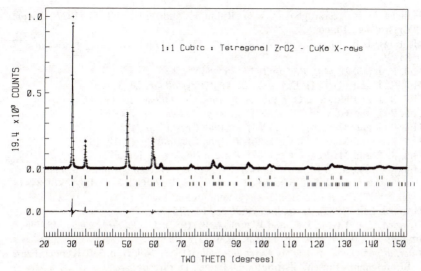

Fig. 5.13 Observed, calculated and difference Cu K_α powder diffraction patterns for a 1:1 by weight mixture of cubic and tetragonal zirconia. The upper and lower rows of tick marks correspond to the positions of all possible reflections from the cubic and tetragonal phases, respectively.

References

Albinati, A. and Willis, B. T. M. (1982). *J. Appl. Crystallogr.*, **15**, 361–74.

Attfield, J. P. (1988) *Acta Crystallogr.*, **B44**, 563–8.

Attfield, J. P., Cheetham, A. K., Cox, D. E., and Sleight, A. W. (1988). *J. Appl. Crystallogr.*, **21**, 452–7.

Bacon, G. E. and Lisher, E. J. (1980). *Acta Crystallogr.*, **B36**, 1908–16.

Baerlocher, C. (1984). *Proceedings of the 6th International Zeolite Conference*, Reno, USA, July 1983. Butterworth Scientific.

Baharie, E. and Pawley, G. S. (1983). *J. Appl. Crystallogr.*, **16**, 404–6.

Baur, W. H. and Fischer, R. X. (1986). *Adv. X-ray Anal.*, **29**, 131–42.

Bendall, P. J., Fitch, A. N., and Fender, B. E. F. (1983). *J. Appl. Crystallogr.*, **16**, 164–70.

Cameron, D. G. and Armstrong, E. E. (1988). *Powder Diffract.*, **3**, 32–8.

Christensen, A. N., Lehmann, M. S., and Nielsen, M. (1985). *Austral. J. Phys.*, **38**, 497–505.

Cooper, M. J. (1983). *Z. Kristallogr.*, **164**, 157–8.

Cooper, M. J., Rouse, K. D., and Sakata, M. (1981). *Z. Kristallogr.*, **164**, 101–17.

David, W. I. and Matthewman, J. C. (1984). Rutherford–Appleton Laboratory Rep. RAL 84-064.

Durbin, J. and Watson, G. S. (1971). *Biometrika*, **58**, 1–19.

Fawcett, T. G., Crowder, C. E., Brownell, S. J., Zhang, Y., Hubbard, C., Schreiner, W. *et al.* (1988). *Powder Diffract.*, **3**, 209–18.

Greaves, C. (1985). *J. Appl. Crystallogr.*, **18**, 48–50.

Hall, M. M. jr., Veeraraghavan, V. G., Rubin, H., and Winchell, P. G. (1977). *J. Appl. Crystallogr.*, **10**, 66–8.

Hastings, J. B., Thomlinson, W., and Cox, D. E. (1984). *J. Appl. Crystallogr.*, **17**, 85–95.

Hill, R. J. (1982). *Mat. Res. Bull.*, **17**, 769–84.

Hill, R. J. and Flack, H. D. (1987). *J. Appl. Crystallogr.*, **20**, 356–61.

Hill, R. J. and Howard, C. J. (1985). *J. Appl. Crystallogr.*, **18**, 173–80.

Hill, R. J. and Howard, C. J. (1987). *J. Appl. Crystallogr.*, **20**, 467–74.

Hill, R. J. and Madsen, I. C. (1984). *J. Appl. Crystallogr.*, **17**, 297–306.

Hill, R. J. and Madsen, I. C. (1986). *J. Appl. Crystallogr.*, **19**, 10–18.

Hill, R. J. and Madsen, I. C. (1987). *Powder Diffract.*, **2**, 146–62.

Hill, R. J., Jessel, A. M. and Madsen, I. C. (1984). *Proceedings of the Symposium on Advances in Lead-Acid Batteries*, New Orleans, USA, Oct. 8–11 1984, Vol. 84–14, pp. 59–77. Electrochemical Society, Pennington, 1984.

Howard, C. J. (1982). *J. Appl. Crystallogr.*, **15**, 615–20.

Howard, C. J., Taylor, J. C., and Waugh, A. B. (1982). *J. Solid State Chem.*, **45**, 396–8.

Immirzi, A. (1980). *Acta Crystallogr.*, **B36**, 2378–85.

International Tables for X-ray Crystallography (1967). Vol. II, p. 330. Kynock Press, Birmingham. (Present distributor D. Reidel, Dordrecht.)

Jansen, E., Schäfer, W., and Will, G. (1988). *J. Appl. Crystallogr.*, **21**, 228–39.

Khattak, C. P. and Cox, D. E. (1977). *J. Appl. Crystallogr.*, **10**, 405–11.

Kisi, E. H., Howard, C. J., and Hill, R. J. (1989). *J. Am. Ceram. Soc.*, **72**, 1757–60.

Klug, H. P. and Alexander, L. E. (1974). *X-ray diffraction procedures for polycrystalline and amorphous materials*. Wiley-Interscience, New York.

Lehmann, M. S., Christensen, A. N., Fjellvåg, H., Feidenhans'l, R., and Nielsen, M. (1987). *J. Appl. Crystallogr.*, **20**, 123–9.

Lewis, J., Schwarzenbach, D., and Flack, H. D. (1982). *Acta Crystallogr.*, **A38**, 733–9.

Louër, D. and Langford, J. I. (1988). *J. Appl. Crystallogr.*, **21**, 430–7.

Maichle, J. K., Ihringer, J., and Prandl, W. (1988). *J. Appl. Crystallogr.*, **21**, 22–7.

McCusker, L. M. (1988). *J. Appl. Crystallogr.*, **21**, 305–10.

Parrish, W., Hart, M., and Huang, T. C. (1986). *J. Appl. Crystallogr.*, **19**, 92–100.

Pawley, G. S. (1980). *J. Appl. Crystallogr.*, **13**, 630–3.

Prince, E. (1985). In *Structure and statistics in crystallography* (ed. A. J. C. Wilson), pp. 95–103, 107. Adenine Press, Guilderland, New York.

Retief, J. J., Engel, D. W., and Boonstra, E. G. (1985). *J. Appl. Crystallogr.*, **18**, 150–5.

Richardson, J. W., Pluth, J. J., and Smith, J. V. (1988). *Acta Crystallogr.*, **B44**, 367–73.

Rollett, J. S. (1982). In *Computational crystallography* (ed. D. Sayre). Clarendon Press, Oxford.

Rollett, J. S. (1988). In *Crystallographic computing 4: new techniques and new technologies* (ed. N. W. Isaacs and M. R. Taylor), pp. 149–66. Oxford University Press/International Union of Crystallography.

Rudolf, P. and Clearfield, A. (1985). *Acta Crystallogr.*, **B41**, 418–25.

Schäfer, W., Jansen, E., Elf, F., and Will, G. (1984). *J. Appl. Crystallogr.*, **17**, 159–66.

Schwarzenbach, D., Abrahams, S. C., Flack, H. D., Gonschorek, W., Hahn, Th., Huml, K. *et al.* (1989). *Acta Crystallogr.*, **A45**, 63–75.

Scott, H. G. (1983). *J. Appl. Crystallogr.*, **16**, 159–63.

Spackman, M. A., Hill, R. J., and Gibbs, G. V. (1987). *Phys. Chem. Minerals*, **14**, 139–50.

Theil, H. and Nagar, A. L. (1961). *J. Am. Stat. Assoc.*, **56**, 793–806.

Thompson, P. and Wood, I. G. (1983). *J. Appl. Crystallogr.*, **16**, 458–72.
Thompson, P., Cox, D. E., and Hastings, J. B. (1987). *J. Appl. Crystallogr.*, **20**, 79–83.
Toraya, H. (1985). *J. Appl. Crystallogr.*, **18**, 351–8.
Von Dreele, R. B., Jorgensen, J. D., and Windsor, C. G. (1982). *J. Appl. Crystallogr.*, **15**, 581–9.
Werner, P.-E., Salomé, S., Malmros, G., and Thomas, J. O. (1979). *J. Appl. Crystallogr.*, **12**, 107–9.
Wertheim, G. K., Butler, M. A., West, K. W., and Buchanan, D. N. E. (1974). *Rev. Sci. Instrum.*, **45**, 1369–71.
Will, G., Masciocchi, N., Parrish, W., and Hart, M. (1987). *J. Appl. Crystallogr.*, **20**, 394–401.
Will, G., Bellotto, M., Parrish, W., and Hart, M. (1988). *J. Appl. Crystallogr.*, **21**, 182–91.
Williams, A., Kwei, G. H., Von Dreele, R. B., Larson, A. C., Raistrick, I. D., and Bish, D. L. (1988). *Phys. Rev. B (Condensed Matter)*, **37**, 7960–2.
Wölfel, E. R. (1983). *J. Appl. Crystallogr.*, **16**, 341–8.
Worlton, T. G., Jorgensen, J. D., Beyerlein, R. A., and Decker, D. L. (1976). *Nucl. Instrum. Methods*, **137**, 331–7.
Young, R. A. and Wiles, D. B. (1982). *J. Appl. Crystallogr.*, **15**, 430–8.
Young, R. A., Prince, E., and Sparks, R. A. (1982). *J. Appl. Crystallogr.*, **15**, 357–9.

5.A Appendix

Quantitative phase analysis

5.A.1 *Introduction*

The presence of multiple phases in a powder diffraction pattern inevitably degrades the resolution of the data and, for the same total counting time, decreases the intensity of the patterns from the individual components. As a result, multiphase samples are an annoying and sometimes pathological problem in crystal structure refinement studies. It has, however, recently become clear that Rietveld analysis of such mixtures can provide very accurate estimates of the relative and/or absolute abundances of the component phases (Werner 1979; Hill and Howard 1987; Bish and Howard 1988; O'Connor and Raven 1988; Hill 1991).

5.A.2 *Analytical relationships*

Quantitative phase, or modal, analysis with the Rietveld method relies on the simple relationship (Hill 1983; Hill and Howard 1987):

$$W_p = S_p(ZMV)_p \bigg/ \sum_{i=1}^{n} S_i(ZMV)_i \qquad (5.A.1)$$

where W is the relative weight fraction of phase p in a mixture of n phases, and S, Z, M, and V are, respectively, the Rietveld scale factor, the number of formula units per unit cell, the mass of the formula unit (in atomic mass units) and the unit cell volume (in $Å^3$). An analogous relationship involving the density of each phase has been presented by Bish and Howard (1988).

If an internal standard phase, s, is added to the mixture in the weight fraction W_s, then the absolute weight fractions of the other identified components p are given by (Howard *et al.* 1988):

$$W_p = W_s S_p (ZMV)_p / S_s (ZMV)_s. \qquad (5.A.2)$$

In this case, a shortfall, from unity, of the sum of the determined weight fractions of the n identified components provides an estimate of the amount of amorphous or non-crystalline material in the sample.

5.A.3 *Advantages of full-pattern analysis*

Application of the Rietveld method to quantitative phase analysis provides many advantages over traditional methods that utilize a small pre-selected set of integrated intensities. In particular:

(1) the calibration constants are computed from simple literature data (i.e. the values of Z, M, and V) rather than by laborious experimentation;

(2) all reflections in the pattern are explicitly included, irrespective of overlap;

(3) the background is better defined since a continuous function is fitted to the whole pattern;

(4) the effects of preferred orientation and extinction are reduced since all reflection types are considered, and in any event, appropriate parameters may often be refined as part of the analysis (see below);

(5) crystal-structural and peak-profile parameters can be refined as part of the same analysis, so that the physical and chemical details of the particular phases in the mixture are adjusted automatically.

5.A.4 *Examples*

Table 5.A.1 shows the results of modal analyses undertaken on various binary mixtures of rutile and corundum using neutron and X-ray data (Hill and Howard 1987; Madsen and Hill 1990) and eqn (5.A.1). The agreement between the experimental and as-weighed compositions is remarkably good for both radiations, although the e.s.d.'s for the more angle-limited X-ray data are noticeably higher than for the corresponding neutron analyses, as expected. Figure 5.A.1 gives the Rietveld analysis plot and phase analysis results for a six-phase mixture in which rutile has been added as an internal

Table 5A.1 Theoretical and experimental phase compositions (wt.%) obtained by Rietveld analysis of X-ray and neutron data collected from mixtures of rutile and corundum

		Measured[a]	
Mixture	As-weighed	X-ray	Neutron
Rutile	70	69.0 (20)	70.5 (3)
Corundum	30	31.0 (12)	29.5 (3)
Rutile	50	48.8 (15)	50.5 (3)
Corundum	50	51.2 (19)	49.5 (3)
Rutile	30	26.5 (7)	30.4 (3)
Corundum	70	73.5 (19)	69.6 (3)

[a] Numbers in parenthesis here, and in the Table 5A.2, represent the error in the least significant figure to the left. The data are from Hill and Howard (1987) and Madsen and Hill (1990).

	Rutile TiO_2	Galena PbS	Pyrite FeS_2	Sphalerite ZnS	Chalcopyrite $CuFeS_2$	Quartz SiO_2	Σ
As weighed (%)	25	15	15	15	15	15	100
Measured (%)	25.0(9)	13.3(12)	13.9(8)	13.8(11)	15.1(8)	14.4(6)	95.5

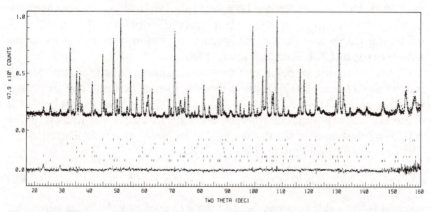

Fig. 5.A.1 Observed, calculated and difference neutron powder diffraction patterns for a synthetic sample consisting of equal proportions of the minerals galena, pyrite, sphalerite, chalcopyrite, and quartz, together with 25 wt.% rutile, added as internal standard. The rows of tick marks, from the top down, indicate the positions of the peaks from all of the component phases, in the order quoted. The results of the quantitative phase analysis obtained by Rietveld analysis of this pattern are shown in the table above the pattern. (From Howard *et al.* 1988.)

standard (Howard *et al.* 1988). In this case, the small short-fall in abundance observed for four of the five 'unknowns', when the data is normalized to the rutile content, suggests the presence of a small amount of surface-amorphous material in these phases.

5.A.5 *Problems*

In general, phase analysis results obtained with X-rays will be inferior to those obtained using neutrons since the difficulties associated with preferred orientation, extinction, micro-absorption, and sample representability are more severe and widespread with the former radiation.

5.A.5.1 *Preferred orientation* Of the problems affecting modal analysis, preferred orientation (PO) is perhaps the most pervasive, and is one which can also have a deleterious effect on the accuracy of the derived crystal structural parameters. Indeed, it has been claimed (Will *et al.* 1983; Will *et al.* 1988) that few powder diffraction patterns are completely free of the effects of PO (nor, incidentally, of the effects of an inadequate 'powder average', with which PO is often confused and incorrectly linked). Thus, it has become routine practice in structure analysis studies (of both the Rietveld and pattern-decomposition kind) to include a correction for PO as part of the refinement procedure.

Most PO corrections involve the approximation of the (assumed single) pole-density distribution by a simple function involving one or two parameters (Rietveld 1969; Toraya and Marumo 1981; Young and Wiles 1982; Will *et al.* 1983; Dollase and Reeder 1986; Valvoda 1987). Other approaches have involved measurement of the pole density distribuions of a number of diffracting planes and their representation by a sum of spherical harmonic terms (Pesonen 1979; Paakkari *et al.* 1988).

Of the relatively popular one-or two-parameter functions, the March distribution (March 1932) displays the best overall performance for crystal structure studies (Dollase 1986; Will *et al.* 1988), has the advantage that it conserves scattering matter (thereby allowing its use in quantitative phase determination), and is equally applicable to platey or acicular crystallites. The pole distribution has the form:

$$P(\alpha) = (r^2 \cos^2 \alpha + \sin^2 \alpha/r)^{-3/2} \qquad (5.A.3)$$

where α is the angle between *hkl* and the PO vector, and r is an adjustable parameter.

The utility and importance of the March correction in quantitative phase analysis has been demonstrated by Rietveld analysis of synthetic mixtures of rutile and corundum, utilizing eqn (5.A.1), as shown in Table 5.A.2. The results obtained using a March correction for both phases (right-hand

Table 5.A.2 Effect of preferred orientation (PO) on the phase compositions (wt.%) determined by Rietveld analysis of X-ray data collected from mixtures of rutile and corundum

As-weighed		No PO correction		PO correction	
Rutile	Corundum	Rutile	Corundum	Rutile	Corundum
70	30	64.5 (14)	35.5 (10)	69.0 (20)	31.0 (12)
50	50	45.4 (11)	54.6 (15)	48.8 (15)	51.2 (18)
30	70	24.8 (5)	75.2 (14)	26.5 (7)	73.5 (19)

PO vectors: (111) for rutile, (104) for corundum.

section of Table 5.A.2) are significantly closer to the as-weighed compositions than those obtained with no correction parameter (central section). As expected, the magnitude of the correction is largest in the case of least dilution of the most oriented phase (rutile).

5.A.5.2 *Extinction* Once thought to be relatively rare in powder diffraction patterns due to the small size of the crystallites, the effects of extinction are now more commonly recognized, especially in phase analysis studies (Cline and Snyder 1987). Sabine (1988; Chapter 4 of this volume) has recently provided a theoretical basis for the general treatment of extinction in polycrystalline materials. Figure 5.A.2 shows the effect of the application of an extinction correction during Rietveld analysis of Cu K_α data collected on a 1:1 by weight mixture of cubic zirconia (c-ZrO_2) with sub-micron-sized corundum (α-Al_2O_3). The scale factors obtained for this mixture gave a 1:1 phase composition only when the c-ZrO_2 component was corrected for extinction effects resulting from the presence of domains of mean size 3.9 μm. It should be noted that this correction is less effectively applied to X-ray data than to neutron data, since it requires a wide angular spread of intense reflections if the effects of extinction are not to be confused with those of the atomic displacement parameters and the pattern scale factors.

5.A.5.3 *Micro-absorption and calibration of the phase analysis* Micro-absorption, like extinction, requires information to be obtained about the particle and domain size distribution of each phase in the sample (Brindley 1945; Hermann and Ermrich 1989; Chapter 4 of this volume). For this reason, and because micro-absorption shows only a small dependence on 2θ, a correction for this effect has not yet been implemented in Rietveld analysis.

However, it is possible to take account of the effects of micro-absorption (and, indeed, any other uncorrected systematic terms) by experimentally

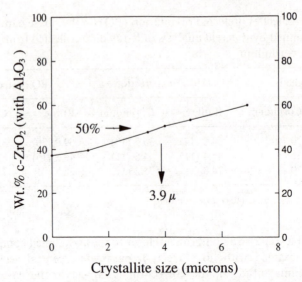

Crystallite size (microns)

Fig. 5.A.2   Phase composition (wt.%) of c-ZrO_2 in a 1:1 mixture with corundum (α-Al_2O_3) plotted as a function of the mean crystallite size (μm) assumed for the c-ZrO_2 component. The modal composition is calculated from the scale factors obtained from a Rietveld analysis of Cu K_α diffraction data, using the 'ZMV' relationship of Hill and Howard (1987), eqn (5.A.1), and the extinction correction of Sabine (1988).

determining the 'constants' in eqn (5.A.1) using a calibration mixture composed of standard materials (Hill 1991). The method is then analogous to the traditional integrated-intensity techniques for phase analysis, but it retains all of the advantages that accrue from consideration of the full-pattern.

5.A.6   *Acknowledgements*

The material in this Appendix was prepared during tenure of a Ludwig Leichhardt Fellowship at the Mineralogisches Institut, Universität Würzburg, Germany; the author is most grateful to the Alexander von Humboldt-Stiftung (Germany) for their generous support.

References

Bish, D. L. and Howard, S. A. (1988). *J. Appl. Crystallogr.*, **21**, 86–91.
Brindley, G. W. (1945). *Phil. Mag.*, **36**, 347–69.
Cline, J. P. and Snyder, R. L. (1987). *Adv. X-ray Anal.*, **30**, 447–56.
Dollase, W. A. (1986). *J. Appl. Crystallogr.*, **19**, 267–72.
Dollase, W. A. and Reeder, R. J. (1986). *Am. Mineral.*, **71**, 163–6.
Hermann, H. and Ermrich, M. (1989). *Powder Diffract.*, **4**, 189–95.
Hill, R. J. (1983). *J. Power Sources*, **9**, 55–71.

Hill, R. J. (1991). *Powder Diffract.*, **6**, 74–7.

Hill, R. J. and Howard, C. J. (1987). *J. Appl. Crystallogr.*, **20**, 467–74.

Howard, C. J., Hill, R. J., and Sufi, M. A. M. (1988). *Chem. Austral.* Oct. 1988, 367–9.

Madsen, I. C. and Hill, R. J. (1990). *Powder Diffract.*, **5**, 195–9.

March, A. (1932). *Z. Kristallogr.*, **81**, 285–97.

O'Connor, B. H. and Raven, M. D. (1988). *Powder Diffract.*, **3**, 2–6.

Paakkari, T., Blomberg, M., Serimaa, R., and Jarvinen, M. (1988). *J. Appl. Crystallogr.*, **21**, 393–7.

Pesonen, A. (1979). *J. Appl. Crystallogr.*, **12**, 460–3.

Rietveld, H. M. (1969). *J. Appl. Crystallogr.*, **2**, 65–71.

Sabine, T. M. (1988). *Acta Crystallogr.*, **A44**, 368–73.

Toraya, H. and Marumo, F. (1981). *Mineral. J.*, **10**, 211–221.

Valvoda, V. (1987). *J. Appl. Crystallogr.*, **20**, 453–6.

Werner, P.-E., Salomé, S., Malmros, G., and Thomas, J. O. (1979). *J. Appl. Crystallogr.*, **12**, 107–9.

Will, G., Parrish, W., and Huang, T. C. (1983). *J. Appl. Crystallogr.*, **16**, 611–22.

Will, G., Bellotto, M., Parrish, W., and Hart, M. (1988). *J. Appl. Crystallogr.*, **21**, 182–91.

Young, R. A. and Wiles, D. B. (1982). *J. Appl. Crystallogr.*, **15**, 430–8.

6

Background modelling in Rietveld analysis

James W. Richardson, Jr

The Rietveld Method was devised as a *complete* powder-diffraction-pattern fitting technique, which necessitates modelling the *total* scattering from crystalline samples. Typical scattering patterns from purely crystalline materials without defects can be reliably modelled simply by convoluting the calculated Bragg intensities with the resolution function of the diffraction instrumentation and adding a background contribution accounting for incoherent scattering, air scattering, and Thermal Diffuse Scattering. (See Chapters 7 and 8 for accounts of crystalline size and microstrain effects on the observed reflection profiles.)

Many diffraction experiments, however, involve the analysis of scattering patterns containing additional non-crystalline scattering components which are not accounted for by conventional background functions. These components are observed in Rietveld analyses as broad oscillations superimposed on the sharp Bragg pattern. One example of this is thermal diffuse scattering (TDS) for which the corrections are well known (Chapter 9; Cooper and Rouse 1968; Windsor 1981). Other examples include scattering from: (1) amorphous containers, (2) incompletely crystallized samples, and (3) separate amorphous phases. These scattering phenomena can be characterized as interference functions developed from short-range interactions between atoms in the sample. Successful refinement of the crystalline structure requires precise modelling of these additional non-crystalline contributions. This can be accomplished through the use of empirical functions such as higher-order polynomials. Alternatively, these contributions can be removed from the overall diffraction pattern while also extracting useful real-space structural information about the non-crystalline scattering. Two such methods are Fourier-filtering (Richardson and Faber 1986) and direct modelling with specially constructed sine series (Larson and Van Dreele 1985–8).

Fourier-filtering involves Fourier transforming residual Rietveld intensities—observed minus (calculated Bragg plus conventional background)—to produce a correlation function related to the radial distribution function (RDF) for the non-crystalline scatterers. Reverse Fourier transformation of the correlation function (summing from $r = 0$ to 5–15, depending on the range of significant correlation) produces a smooth fit to the oscillatory component which can be subtracted from the original data for continued refinement. Direct modelling of the oscillatory component involves representing the correlation function in the above reverse Fourier transformation as a set of delta-functions with variable position and height. In both cases maxima in the correlation correspond to interatomic distances prevalent in the non-crystalline component.

In typical Rietveld refinements of purely crystalline materials, the observed diffraction intensities $y_{obs}(Q)$ can be modelled by summing the contributions from crystalline Bragg scattering and background scattering:

$$y_{obs}(Q) = y_{cx}(Q) + y_b(Q) \tag{6.1}$$

where $y_{cx}(Q)$ and $y_b(Q)$ are the calculated crystalline and background intensities, respectively, and $Q = 4\pi \sin \theta/\lambda$ is the magnitude of the scattering vector. Reciprocal spacings are given in units of Q instead of $\sin \theta/\lambda$ for simplicity of calculation in Fourier sums (see below). The background is commonly fitted with low-order polynomial functions. In some cases, however, there are contributions from short- to intermediate-range-order scattering where conventional background functions are unable to account for all non-crystalline scattering, i.e.

$$y_{obs}(Q) = [y_{cx}(Q) + y_b(Q)] + y_{ca}(Q) \tag{6.2}$$

where $y_{ca}(Q)$ now are the additional non-crystalline intensity contributions. Depending on the correlation lengths involved, the sharpness of features in $y_{ca}(Q)$ can begin to approach that of the Bragg peaks, so care must be taken to obtain values for $y_{ca}(Q)$ which are independent of the crystalline refinement. Higher-order polynomials can be used to estimate $y_{ca}(Q)$, but the extent of the corrections is difficult to control. In addition, this gives us no new insights into the atomic entities giving rise to the non-crystalline scattering. Treating the additional scattering as though it were from an amorphous material (which it often is) *can* provide additional structural information of some value.

A case in point is $AlPO_4$-5, an aluminophosphate framework molecular sieve. Figure 6.1 shows the Rietveld refinement plot in which observed neutron powder diffraction intensities ('+') are plotted along with the sum of calculated crystalline and conventional background intensities (smooth curve). The residual at the bottom shows evidence for additional scattering, which might be suspected to arise from an amorphous component in the sample. The amorphous scattering is clearly quite prominent. In fact, extending the range (Fig. 6.2), reveals that the dominant scattering at high Q (small d-spacing) is from this additional component.

Treating this scattering as amorphous, we can calculate the correlation function, $d(r)$ from the numerical approximation to the 1-dimensional

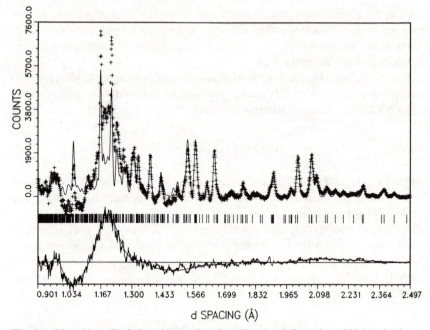

Fig. 6.1 Rietveld profile fit for Aluminophosphate Number 5, from the ±90° data banks on the GPPD time-of-flight diffractometer, before Fourier-filtering. The plus signs (+) are the observed, background subtracted intensities, $y_{obs}(Q) - y_b(Q)$. The solid line represents the calculated crystalline intensities, $y_{cx}(Q)$. The residual intensities are shown at the bottom of the figure (jagged line), along with the calculated non-crystalline intensity contribution (smooth line). Tick marks below the profile indicate the positions of the Bragg reflections.

Fourier transform of the quantity $[S(Q) - 1]$:

$$d(r) = \frac{2}{\pi} \sum_{Q_{min}}^{Q_{max}} QM(Q)[S(Q) - 1] \sin(Qr) \, \Delta Q \qquad (6.3)$$

where

$$[S(Q) - 1] = [y_{obs}(Q) - y_{cx}(Q) - y_b(Q)]/y_{inc}(Q), \qquad (6.4)$$

$M(Q)$ is a modification function commonly used in amorphous diffraction analyses:

$$M(Q) = \sin(\pi Q/Q_{max})/(\pi Q/Q_{max}) \qquad (6.5)$$

and $y_{inc}(Q)$ is the intensity in the incident spectrum (not flat for time-of-flight neutron diffraction analysis). The correlation function $d(r)$ contains maxima at interatomic spacings characteristic of the short-range interactions giving rise to non-crystalline scattering.

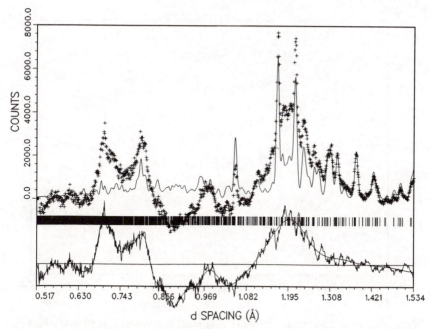

Fig. 6.2 Profile plot for Aluminophosphate Number 5 as in Fig. 6.1, with a lower minimum d-spacing. Note that the non-crystalline component dominates the crystalline.

Figure 6.3 shows d(r) for the amorphous component in the AlPO$_4$-5 sample. The positions of many of the observed peaks are close to interatomic distances also characteristic of the crystalline material. The potential accuracy of the structural information available in d(r) is summarized in Table 6.1. The close agreement suggests that the amorphous component, although lacking long-range periodicity, has an atomic arrangement similar to that of the crystalline material.

In Fourier-filtering a smoothed $y_{ca}(Q)$ is calculated from the reverse Fourier summation of d(r) values:

$$y_{ca}(Q) = y_{inc}(Q) \sum_{r_{min}}^{r_{max}} d(r) \frac{\sin(Qr)}{(Qr)} \Delta r \qquad (6.6)$$

Filtering is accomplished by setting r_{max} at an appropriate value ranging from ~5.0–15.0, depending on the range of significant correlation. For AlPO$_4$-5, a correlation length of 15.0 was required to account for features clearly due to the amorphous component. The smoothed $y_{ca}(Q)$ (shown in Fig. 6.2 superimposed on the residual Rietveld intensity) can then be subtracted from the original raw data and the Rietveld refinement continues.

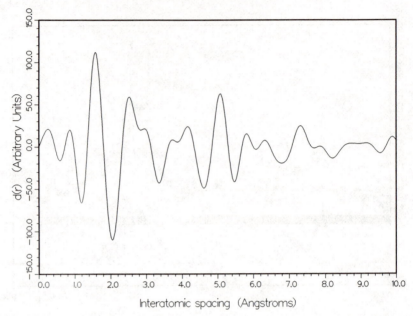

Fig. 6.3 Correlation function d(*r*) calculated using residual intensities from Rietveld refinement of Aluminophosphate Number 5.

Table 6.1 Observed interatomic spacings in crystalline and non-crystalline components of $AlPO_4$-5

Atoms	Neighbour	Crystalline	Non-crystalline
T–O[a]	1st	1.6	1.6
O–O	2nd	2.6	2.5
T–T′[b]	2nd	3.1	3.0
O–O	3rd	3.7	3.7
T–T″[c] (4-ring)	4th	4.4	4.2
T–T‴[c] (6-ring)	4th	5.4	5.1

[a] T corresponds to Al and P which are indistinguishable in this analysis.
[b] T–T′ corresponds to distance between neighbouring tetrahedral centres.
[c] T–T″ and T–T‴ correspond to distances between 2nd neighbour tetrahedral centres, e.g. across rings.

For statistical accuracy, the Rietveld program (Rotella 1988) in which the Fourier-filtering is being implemented calculates σ for each data point based on the original intensities, not corrected intensities. Furthermore, corrections are made on the raw data only; no corrections on corrections. Because the Rietveld refinement, when working with uncorrected data, will tend to

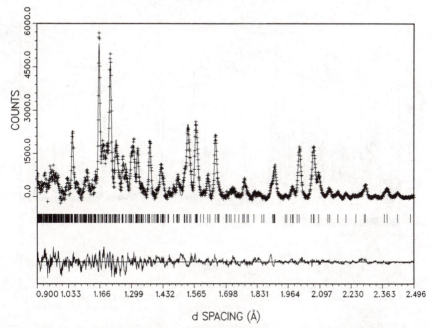

Fig. 6.4 Rietveld profile fit for Aluminophosphate Number 5 after three iterations of Fourier-filtering to remove non-crystalline scattering component.

compensate for non-crystalline scattering with thermal parameter adjustments etc., more than one iteration is usually needed to completely and accurately model the scattering. Figure 6.4 shows the result of three iterations of Fourier-filtering on $AlPO_4$-5.

Direct modelling of an amorphous component in a crystalline sample would entail parameterizing the correlation function d(r) such that eqn (6.6) can be rewritten

$$y_b(Q) = B_1 + B_2/Q + \sum_{i=1}^{5} B_{2i+1} \sin(QB_{2i+2})/(QB_{2i+2}) \qquad (6.7)$$

where B_1–B_{12} are variable parameters. In this formalism, d(r) is expressed as a set of five delta-functions positioned at maxima (B_{2i+2}) in d(r) with heights (B_{2i+1}) equal to the magnitude of d(r) at B_{2i+2}.

What are the consequences to the Rietveld refinement? This question can be answered with a calibration experiment. The Rietveld plot for a mixed sample of 50 per cent (by weight) crystalline quartz and 50 per cent amorphous silica is shown in Fig. 6.5. The scattering from the silica is immediately obvious and can be removed (Fig. 6.6). The crystalline structural

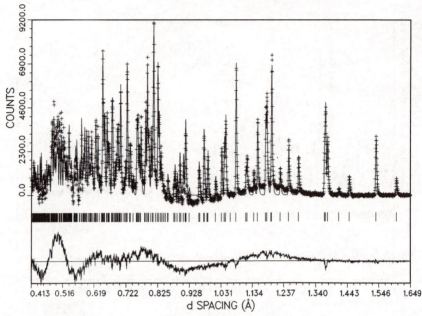

Fig. 6.5 Rietveld profile fit for a 50:50 mixture of crystalline quartz and amorphous silica, before Fourier-filtering. See Fig. 6.1 for explanation of symbols.

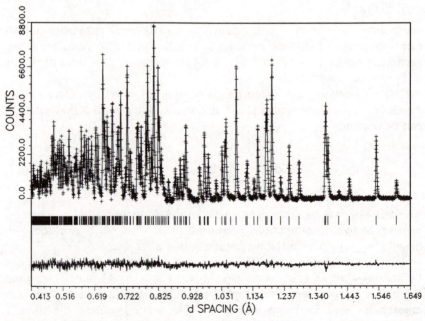

Fig. 6.6 Result of Fourier-filtering for 50:50 mixture of crystalline and amorphous silica.

Table 6.2 Comparative refinements of quartz with and without silica added

Parameter	Mixture[a]	Quartz[b]	X-ray single-crystal[c]
Si, x	0.4698 (2)	0.4700 (2)	0.4698 (1)
U_{11}	0.0059 (3)	0.0062 (3)	0.0070 (1)
U_{22}	0.0050 (4)	0.0050 (6)	0.0054 (1)
U_{33}	0.0053 (3)	0.0065 (9)	0.0061 (1)
U_{13}	0.0000 (1)	0.0007 (2)	0.0001 (1)
O, x	0.4140 (2)	0.4131 (2)	0.4137 (2)
y	0.2675 (1)	0.2677 (2)	0.2677 (2)
z	0.2143 (1)	0.2144 (1)	0.2145 (1)
U_{11}	0.0141 (3)	0.0179 (4)	0.0154 (3)
U_{22}	0.0100 (3)	0.0110 (3)	0.0111 (3)
U_{33}	0.0102 (1)	0.0107 (1)	0.0113 (2)
U_{12}	0.0078 (3)	0.0095 (3)	0.0088 (2)
U_{13}	0.0027 (2)	0.0027 (2)	0.0030 (2)
U_{23}	0.0040 (1)	0.0052 (2)	0.0046 (2)
a, b, c	4.9137 (1)	4.9141 (1)	4.9138
	5.4053 (1)	5.4060 (1)	5.4054

[a] Quartz-silica mixture, after Fourier-filtering.
[b] Earlier refinement with neutron powder diffraction data: Lager *et al.* (1982).
[c] Single-crystal X-ray refinement: Le Page *et al.* (1979).

parameters obtained from data corrected using Fourier-filtering are compared with a previous neutron powder diffraction refinement of quartz and recent single-crystal X-ray results in Table 6.2. All refined parameters agree within 3–4 σ.

Conclusion

There is, indeed, additional information available in 'background' scattering. Fourier-filtering allows the user to control the background correction in a physically meaningful way, namely by adjusting r_{max} in eqn (6.6). Finally, it is important to point out that removing 'problem' scattering will never provide you with better results than having a 'clean' sample to begin with. This work was supported by the US Department of Energy under Contract No. W-31-109-ENG-38.

References

Cooper, M. J. and Rouse, K. D. (1968). *Acta Crystallogr.*, **A24**, 405–12.
Lager, G. A., Jorgensen, J. D., and Rotella, F. J. (1982). *Appl. Phys.*, **53**, 10.

Larson, A. C. and von Dreele, R. B. (1985–8). GSAS: Generalized structure analysis system. Los Alamos National Laboratory, USA, LAUR 86-748.

Le Page, Y., Calvert, L. D., and Grabe, E. J. (1979). *J. Phys. Chem. Solids*, **41**, 721.

Richardson, J. W. and Faber, J. jr (1986). *Adv. X-ray Anal.*, **29**, 143–52.

Rotella, F. J. (1988). *Users manual for Rietveld analysis of time-of-flight neutron powder diffraction data at IPNS*. Argonne National Laboratory, USA.

Windsor, C. G. (1981). *Pulsed neutron scattering*, pp. 294–5. Halsted Press, New York.

7

Analytical profile fitting of X-ray powder diffraction profiles in Rietveld analysis

Robert L. Snyder

7.1 Introduction

The inherent asymmetry of many powder diffraction profiles has been a principal hindrance in extending the application of diffraction techniques. The resurgence of developments in this area in recent years is primarily the result of the availability of general algorithms permitting computer automation of powder diffractometers (Mallory and Snyder 1979). The ability to collect digitized representations of the line profiles and apply numerical methods to their analysis has led to a number of new and exciting applications (Snyder 1983).

The most exciting of these has been the continuing development of the whole pattern X-ray refinement method to obtain crystal structure information, begun by Rietveld (1969), who applied it to neutron diffraction (Chapter 2). The mostly symmetric-Gaussian nature of neutron diffraction lines has aided this application of the method to develop to its current rather mature state.

The application of the Rietveld method to X-ray patterns has been slower to develop, primarily because of the asymmetric and non-Gaussian nature of, and multiple spectral components in, most X-ray diffraction profiles. Malmros and Thomas (1977) and Young *et al.* (1977) gave the first applications of this technique to X-ray data and the work of Wiles and Young (1981) marks the beginning of the much wider development of this method. However, the fact remains that in many current refinements, a major portion of the least squares residual is determined by the profile mismatch between the model and observed data rather than being determined by the fit of the structural parameters. The various successes fitting asymmetric profiles with symmetric functions, reported to date, have often been the result of defect disorder which tends to obliterate the inherent peak asymmetry and produce a more symmetric shape. Most current workers have not allowed separately for the profile shape function contributions from either the instrument or the sample.

7.2   The origin of the profile shape

There once was a time when a discussion of the shape of X-ray powder diffraction profiles could have been happily limited to the effects seen using a Bragg–Brentano geometry diffractometer with a sealed tube source. In fact this has been done very well by Klug and Alexander (1974). Today a broader discussion is required to cover the common use of both pulsed and steady state neutrons and of synchrotron and rotating anode X-ray sources with incident and/or diffracted beam monochromators. Even the advances in X-ray detectors have an impact on this topic.

A diffraction line profile is the result of the convolution of a number of independent contributing shapes, some symmetric and some asymmetric. The process of convolution is one in which the product of two functions is integrated over all space. As is noted in Chapter 1 (eqn 1.7), it can be represented as:

$$h_{2\theta} = g_{2\theta} * f_{2\theta} = \int_{-\infty}^{+\infty} g_{2\theta'} f_{2\theta - 2\theta'} \, d(2\theta') \qquad (7.1)$$

where $h_{2\theta}$ is the final observed profile and $g_{2\theta'}$ and $f_{2\theta}$ are shape functions contributing to the resulting profile. Each point in the convolution is the result of summing the product of g and f over all possible values for $f_{2\theta}$. It is clear that, if this operation cannot be performed analytically, a lot of computer time will be used to evaluate it numerically. However, this may be reduced by the proper choice of functions or by carrying it out in Fourier space.

The components contributing to a diffraction profile can be divided into three categories.

7.2.1   Intrinsic profile: f

The dynamical diffraction of an X-ray beam in a perfect crystal produces a reflection whose inherent width is called the Darwin width, after the author of the first dynamical treatment of diffraction. This inherent width is simply the result of the uncertainty principle ($\Delta p \Delta x = h$), in that the absorption coefficient of the specimen requires that the location of the photon in a crystal is restricted to a rather small volume. This means that Δp and in turn $\Delta\lambda$ ($\Delta p = h/\Delta\lambda$ by the de Broglie relation) must be finite, producing a finite width to a diffraction peak (Sabine 1989). The Darwin profile has sometimes been represented by a Lorentzian function (Parrish *et al.* 1976).

In addition to the inherent width, there are two principal physical sample effects which will broaden the profile shape function, f, that the specimen contributes to the observed profile. Scherrer first pointed out that as crystallite size decreased below about 1 μm, the integral breadth β, of the profile would increase according to

$$\beta = \frac{\lambda}{\tau \cos \theta} \tag{7.2}$$

where τ is the 'X-ray crystallite size'.

Microstrain also broadens the specimen profile, f, according to

$$\beta = k\varepsilon \tan \theta \tag{7.3}$$

where ε represents the microstrain and k is a constant whose value depends on the definition of microstrain used. See Langford *et al.* (1988) and Delhez *et al.* (1988 and Chapter 8) for recent discussions of these effects. Both of these specimen broadening effects are generally modelled as being symmetric but microstrain broadening need not be. Compositional inhomogeneities might be one of a number of possible sources of broadening that can also vary as tan θ (Chapter 8; Langford and Louër 1991).

7.2.2 *Spectral distribution: W*

The most common X-ray source continues to be the sealed target X-ray tube. The inherent spectral profile of the $K_{\alpha 1}$ line from a Cu target X-ray tube has a breadth of 0.518×10^{-3} Å (Edwards and Langford 1971) and has been shown to be approximately Lorentzian and not completely symmetric—see Frevel (1987) for a recent discussion. The inherent width and asymmetry is usually overwhelmed by the fact that the various components of radiation ($K_{\alpha 1}$, $K_{\alpha 2}$, and $K_{\alpha 3,4}$) in a polychromatic beam will each spread out as 2θ increases (i.e. with tan θ). This spectral dispersion is so great that it can dominate the diffraction profiles at high angle, making them quite broad and, except for the effects of axial divergence, relatively symmetric.

Monochromatization of an X-ray or neutron beam using an incident beam monochromator limits the breadth of wavelength profile function W, to the Darwin width of the monochromator crystal and its mosaicity. The use of both an incident and a diffracted beam monochromator produces broadening effects that depend on the displacement angle of a diffraction profile away from the point of optimum focus, which in turn depends on the Bragg reflections chosen for the monochromator and analyser crystals (Cox *et al.* 1988). At the point of optimum focus, using perfect germanium crystals with a synchrotron source, the instrumental profile is so narrow that the observed breadth is due primarily to the collimator divergence.

7.2.3 *Instrumental contributions: G*

There are five principal non-spectral contributions to the instrumental profile, depending on the instrumental arrangement.

1. The X-ray source image. In a closed tube system this can be approximated with a symmetric Gaussian curve with a full-width-at-half-maximum (Γ) of 0.02° using a take-off angle of 3°. The effects of incident and diffracted beam monochromators are described by Cox *et al.* (1988) and do not introduce any asymmetry to the Gaussian shape. The long beam lengths and small effective source sizes at synchrotron ports allow for nearly perfect parallel optics. This, when coupled with both an incident and diffracted beam monochromator leaves the instrumental profile, at optimum focus, nearly a delta-function (Cox *et al.* 1983). The use of focusing (curved crystal) optics with an incident beam slit following the monochromator, while greatly increasing the intensity, introduces significant but symmetric broadening (Parrish *et al.* 1986).

2. Flat specimen. To maintain the Bragg–Brentano focusing condition the sample should be curved so that it follows the focusing circle. Since the focusing circle continuously changes radius with 2θ, most experimental arrangements simply use a flat specimen, tangent to the focusing circle. This 'out of focus' ('parafocusing') condition introduces a cot θ dependence and produces a small asymmetry in the profile. It is particularly noticeable at low angles, where the irradiated length of the sample is large. This term is not present on those neutron and synchrotron devices that use a cylindrical sample bathed in the beam.

3. Axial divergence of the incident beam. This follows an approximate cot θ dependence at low angle and causes a substantial asymmetry in the profile, particularly at the lowest angles (Wilson 1963).

4. Specimen transparency. As the absorption coefficient of the sample decreases, the X-ray beam penetrates ever deeper, making the effective diffracting surface farther and farther off from the focusing circle. This produces a substantial asymmetric convolution term for low μ materials.

5. Receiving slit. For instruments using a receiving slit, another symmetric term contributes to the observed profile.

7.2.4 Observed profile: $h(x)$

Each of the terms described above conspires, via convolution, to produce a final diffraction profile that will range from being very asymmetric in the case of sealed tube Bragg–Brentano systems to quite symmetric, nearly Gaussian profiles, in the cases of fixed energy neutron and some synchrotron X-ray instruments.

The observed diffraction profile, as stated by Jones (1938) and applied to powder diffraction systems by Taupin (1973) and Parrish *et al.* (1976) is the result of the convolution of a specimen profile (f) and a combined function modelling the aberrations introduced by the diffractometer and wavelength dispersion. Taupin and Parrish grouped these terms together as $W * G$. The overall line profile can be expressed as:

$$h(x) = (W * G) * f(x) + \text{background} \qquad (7.4)$$

where the ($*$) represents the convolution operation and x is either an angle or an energy variable (see discussion of eqn 1.7 in Chapter 1). Since both W and G are fixed for a particular instrument/target system, $(W * G)$ may be regarded as a single entity which we will refer to as the instrumental profile $g(x)$. The specimen function $f(x)$ for a sample with no defect broadening has only the Darwin width which can be approximated with a delta-function (i.e. a profile of infinite height and zero width). Using a delta function for $f(x)$ in eqn (7.4) yields:

$$h(x) = g(x) + \text{background}. \qquad (7.5)$$

Hence, for a pattern of an ideal sample, with background approximately accounted for, the profiles are identical to the profiles of $g(x)$. However, Parrish *et al.* (1976) and Howard and Snyder (1983) have indicated that the intensity ratio of components is affected by the setting of the monochromator, when one is present, and hence the W component of $g(x)$ must also be evaluated.

7.3 Modelling of profiles

Khattak and Cox (1977) have shown fundamental problems in representing X-ray diffraction lines with either the Gaussian or simple Lorentzian functions. Of more recent interest are the Voigt (Langford 1978; Cox *et al.* 1988), the pseudo-Voigt (see Young and Wiles 1982) and the split-Pearson VII of Brown and Edmonds (1980). The Voigt function is the result of an analytical convolution of a Gaussian and a Lorentzian. It therefore ranges from pure Lorentzian to Gaussian type, depending on the ratio of the component widths. The function normalized to 1.0 has the form

$$V(x, \Gamma_L, \Gamma_G) = L(x/\Gamma_L) * G(x/\Gamma_G) \tag{7.6}$$

where, L = Lorentzian function with a full-width-at-half-maximum (FWHM) of Γ_L, G = Gaussian function with FWHM of Γ_G, and $x = \Delta 2\theta$. The Voigt-function can be calculated numerically using the complex error function.

The pseudo-Voigt conveniently allows the refinement of a mixing parameter determining the fraction of Lorentzian and Gaussian components needed to fit an observed profile. See Hastings *et al.* (1984), Cox *et al.* (1988) and Chapters 8 and 9 for applications of this function. The pseudo-Voigt, although symmetric, allows for a flexible variation of the two most common profiles ranging from the broad Γ Gaussian to the narrow Γ Lorentzian. However, Smith *et al.* (1987) have found an example of synchrotron profiles which cannot be fitted by the Voigt function because their shapes lie outside the Gaussian-to-Lorentzian range (see also Wertheim *et al.* 1974).

The Pearson VII function also allows for the variation in shape from Lorentzian to Gaussian (Howard and Snyder 1983) and is also symmetric. In our previous work (Howard and Snyder 1983, 1985*a*, 1989), we evaluated seven profile models with three regression techniques and concluded that the split-Pearson VII function of Brown and Edmonds (1980), combined with a Gauss–Newton or a Marquardt least-squares optimization algorithm, gave excellent fits to asymmetric X-ray diffraction lines obtained under a wide variety of conditions. The approach here is to split each diffraction profile into a low angle and high angle part by dividing at the profile maximum, and fit a separate Pearson VII to each side. The Pearson VII function has the form (cf. Table 1.2 in Chapter 1):

$$I = I_0/(1 + ax^2)^m \tag{7.7}$$

where $a = (2^{1/m} - 1)/(\Gamma/2)^2$, m is the shape factor whose value determines the rate at which the tails fall and Γ is the full-width-at-half-maximum. The split-Pearson VII, as illustrated in Fig. 7.1 was shown consistently to refine against observed profiles with the lowest residual error of the seven models tested. The principal problem with this function is that its clumsy form prevents the analytical determination of convolutions and thus forces extensive numerical approximations.

There are several demands on a profile shape function.

- The function must fit non-symmetric peaks.

- It should be mathematically as simple as possible, to make possible the calculation of all derivatives to the variables, and

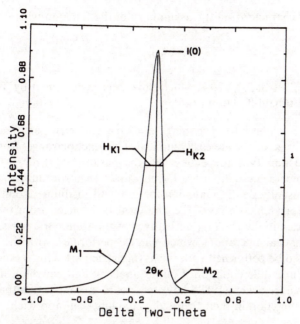

Fig. 7.1 The split-Pearson VII function uses two halves of the Pearson VII function with a common peak angle and intensity. The breadth Γ and the exponents m for each side are varied independently.

- allow simple computation of the integral intensity.

- The convolution with a Lorentzian or a Gaussian function modelling f should be possible analytically.

Tomandl (1987) has proposed the following profile shape function:

$$T(2\theta) = \eta[L(x) - asxL^2(a_L x)] + (1 - \eta)[G(x) - as(x + x_0)G(a_G(x + x_0))]$$

$$(7.8)$$

where,

$L(x)$ = Lorentzian function,

$G(x)$ = Gaussian function,

$x = (2\theta - 2\theta_{peak})/\Gamma$,

2θ = peak position,

Γ = full-width-at-half-maximum,

η = mixing factor, giving the proportion of the Lorentzian part,

as = factor characterizing the asymmetry of the function where

$$0 < |as| < 0.5,$$

$a_L = 0.6$, $a_G = 0.85$, $x_0 = 0.7$, which have been determined by Tomandl to fit sealed tube diffraction profiles.

The function is split into a symmetric and an asymmetric part. The first three conditions listed above are easily fulfilled. The function ranges from Lorentzian to Gaussian type (dependent on the parameter η), describing X-ray profiles, which are usually more Lorentzian, to neutron diffraction profiles, which are usually nearly Gaussian for low and medium resolution instruments. The asymmetric parts (i.e. the terms subtracted from the L and G parts) are essentially the first derivatives of the symmetric parts. Therefore all mathematical operations which can be performed with the symmetric part, can also be performed with the asymmetric part. This is especially true for integration, differentiation, Fourier analysis, and convolution. Recently Hepp and Baerlocher (1988) have proposed a similar treatment for an asymmetry function in their 'learned shape function' approach.

7.4 The X-ray spectrum

Let us now examine in detail a procedure for fitting X-ray diffractometer profiles using split Pearson functions, following that of Howard and Snyder (1983, 1985b, 1989). The spectrum from a Cu X-ray tube target has the $K_{\alpha1}$, $K_{\alpha2}$, and $K_{\alpha3,4}$ wavelengths as the dominant components. We may take advantage of knowing their respective wavelengths and relative intensities to create a 'compound profile' by calculating the position and intensities for the diffracted lines for the $K_{\alpha2}$ and $K_{\alpha3,4}$ based on the parameters of the $K_{\alpha1}$. If the profile shape function has these three components 'built-in', then its shape will allow it to fit only those areas of the pattern that exhibit the intrinsic $K_{\alpha1}$, $K_{\alpha2}$, $K_{\alpha3,4}$ triplet. Constraining the $K_{\alpha2}$ and $K_{\alpha3,4}$ components in this manner greatly reduces the number of variables in the refinement and thereby lowers the number of false minima likely to slow down or entrap the Rietveld refinement process.

The fixing of the $K_{\alpha2}$ and $K_{\alpha3,4}$ peak positions and intensities in the split-Pearson VII function reduces the number of parameters in the compound profile function from 18 to 14, as is illustrated in Fig. 7.2. However, the independent variation of the full-widths-at-half-maximum, Γ, and exponential shape factors, m, of the $K_{\alpha2}$ and $K_{\alpha3,4}$ lines pose a problem, particularly when the lines are not resolved. Refinement in this case may show parameter values oscillating and not converging, or refined parameters that physically do not make sense (e.g. strongly distorted profile shapes).

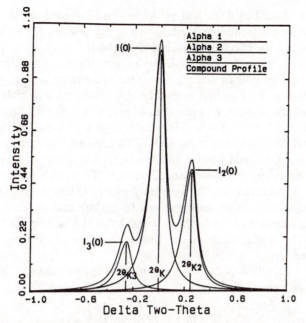

Fig. 7.2 The compound split-Pearson VII profile shape function with the $K_{\alpha2}$ and $K_{\alpha3,4}$ components 'built-in'. (The $K_{\alpha3}$ is exaggerated by a factor of 20 for clarity). For a particular set of $2\theta_k$ and I_0 values, $2\theta_{K2}$, $2\theta_{K3}$, I_{02}, and I_{03} can be calculated. This leaves six values for m and six for Γ_K or 14 total to be determined.

To eliminate a few parameters, while minimizing the effects due to approximations, the shape factors, m, and the Γ's of the weak $K_{\alpha3,4}$ are fixed to be the same as the $K_{\alpha1}$. This reduces the number of parameters to 10. Ladell *et al.* (1970) have shown that the wavelength dispersion difference between the $K_{\alpha1}$ and $K_{\alpha2}$ causes their profiles to be slightly different. This small profile shape difference is ignored in constructing the compound profile for the split-Pearson VII.

7.5 Description of background

The description of the background in a powder diffraction pattern is critical to profile shape and area refinement because any background function must correlate strongly with the profile function, see for example, Chapters 4, 6, and 9 by Sabine, Richardson, and Suortti, respectively, in this book. Two commonly used methods to describe background involve (i) selecting points between peaks and interpolating between them (Rietveld 1969) and (ii) refining the coefficients of a polynomial along the profile parameters (Wiles and Young 1981). While Sabine (1977) has described an analytical method for neutron powder diffraction by which the background scattering is analytically described from first principles and included in the refinement, there is no simple technique available for X-ray powder diffraction (but see Chapter 6 for an effective non-simple method). In addition, no method proposed to date (Snyder 1983) will routinely produce an accurate description of X-ray background in complex patterns. The most serious problem in finding the true background occurs in low symmetry materials where the powder pattern presents a continuum of peaks at high diffraction angle.

Howard and Snyder (1985a, 1989) incorporated an empirical approach to background determination in the SHADOW algorithm (1985b). Background was carefully measured from a silicon crystal cut along a plane for which the Bragg reflection was systematically absent. Thus neither amorphous nor crystalline scattering contributions are present. Although there is no theoretical reason for sample contributions to background to cause a uniform vertical displacement of this function, we find that this assumption 'adequately' described the background in the limited number of diffraction patterns investigated.

Visual examination of the silicon pattern at lowest values of 2θ suggested a functional form of the type:

$$I_{(i)} = A2\theta_{(i)} + B(2\theta_{(i)} - C)^D + O \tag{7.9}$$

where A, B, C, and D are constants obtained from a least squares fit to a quartz pattern. The offset O is the only background variable used during profile refinement. This approach greatly minimizes the correlations between profile and background parameters during profile refinement.

7.6 Unconstrained profile fitting

The SHADOW algorithm (Howard and Snyder 1985b) was developed in FORTRAN 77 as a general profile or pattern fitting program which permits the use of a wide variety of shape functions and either a Gauss–Newton or Marquardt optimization algorithm. It permits the use of different profile models ranging from a simple Lorentzian or Pearson VII function to the compound split Pearson required to model $K_{\alpha 2}$ and $K_{\alpha 3,4}$ components. Thus it has been used to fit the very narrow lines from synchrotron sources, through incident beam monochromated sealed tube systems to conventional graphite diffracted beam monochromated systems with all α components present. As the complexity of the required profile model increases, the number of adjustable parameters must also increase.

Figure 7.3 shows a simple unconstrained split-Pearson function fitting two peaks measured with a diffractometer using an incident beam monochromator. The fit is quite good with the individual profiles looking like realistic diffraction lines. Figure 7.4 shows the same two peaks measured on

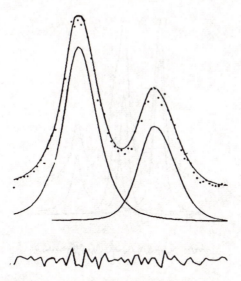

Fig. 7.3 Correct profile fit of a doublet measured with only $K_{\alpha 1}$ radiation, using a split-Pearson function.

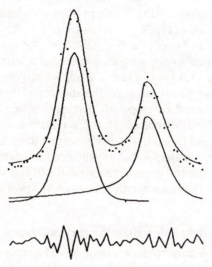

Fig. 7.4 Correct profile fit of a doublet measured with $K_{\alpha 1}$ and $K_{\alpha 2}$, using a split-Pearson function.

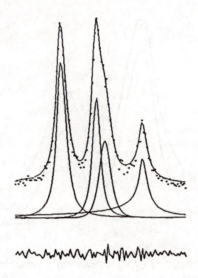

Fig. 7.5 Incorrect profile fit of a doublet measured with only $K_{\alpha 1}$ radiation, using a split-Pearson function.

a similar diffractometer but with different resolving power. Here we see that the low angle side of the split-Pearson fitting the peak on the right has lifted in an unrealistic manner. Figures 7.5 and 7.6 show the same peak group measured on two instruments without incident beam monochromators and illustrate a fundamental problem of profile fitting. While the difference curves indicate a good fit, some of the fitting profiles are clearly unrealistic. Note that the low angle side of the $K_{\alpha 2}$ profile in Fig. 7.5 lifts unrealistically while the rest of the profiles fit well. In Fig. 7.6 we see the low angle sides of two split-Pearsons lift improperly, yet the difference plot looks very good. In fact there are a number of ways to place a group of profiles under an envelope if the only criterion is to minimize the differences between the summed profiles and the observed envelope. All that any optimization procedure can do is to minimize these differences. Thus it is clear that unconstrained profile fitting can always lead to failure to give a correct fit unless we constrain the profiles used to be of the shapes that are allowed for a particular experimental configuration. Any general procedure for profile fitting must, then, allow for the application of experimental constraints on the profile shape.

7.7 Establishing profile constraints: the $g(x)$ calibration curve

In order to establish constraints on the allowed shapes of the profiles on a particular instrument, we determine the instrument $g(x)$ function as described by Parrish *et al.* (1976). The profiles of a standard which shows no sample

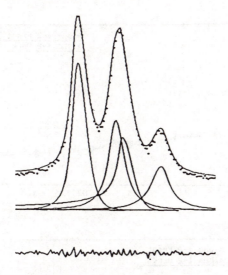

Fig. 7.6 Incorrect profile fit of a doublet measured with $K_{\alpha 1}$ and $K_{\alpha 2}$, using a split-Pearson function.

broadening are fitted with the split-Pearson VII function. Ideally one would like a uniform crystallite size of about 5 µm, for an unstrained, high-symmetry material to use as a standard. Although the US National Bureau of Standards SRM640A Si sample has a small amount of crystallite size broadening, it was chosen as the standard for want of a better material at the time. SRM640B also shows some size broadening. The National Institute for Science and Technology (formerly NBS) has recently released a LaB_6 standard reference material for instrumental profile calibration.

The 11 profiles of Si, observed with Cu K_{α} radiation, were fitted with the constrained split-Pearson function, described above. Each profile was refined separately in a region with enough points to allow a precise description of the profile. In each case the background parameters were refined along with the profile parameters.

The Γ values for each Pearson VII component obtained from the refinement were used to determine the value of the coefficients in the polynomial expression derived for neutron diffraction by Caglioti *et al.* (1958):

$$\Gamma_K^2(2\theta) = U \tan^2(\theta_K) + V \tan(\theta_K) + W. \tag{7.10}$$

An equation of this type was established by least-squares analysis of the parameters obtained from the low angle side of the split-Pearson VII functions, which had been used to analyse the profiles from the Si standard. A second equation was obtained for the high angle sides of the standard

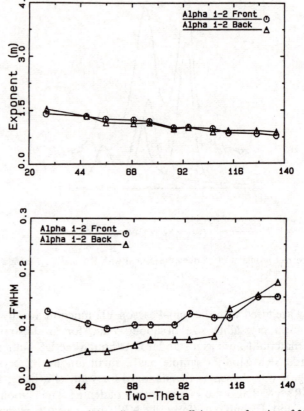

Fig. 7.7 Variation of the split-Pearson coefficients as a function of 2θ.

profiles. Similarly, a polynomial was used for the shape factors, m:

$$m = a'(2\theta_K)^2 + b'(2\theta_K) + c'. \tag{7.11}$$

Least squares regression of this function versus the two sets of shape factors completes the establishment of an analytical expression for evaluating $g(x)$. Example polynomials for the high and low angle split-Pearson parameters (m and Γ) for the calibration of a conventional diffractometer are illustrated in Fig. 7.7. This calibration procedure allows the evaluation of the instrumental $g(x)$ profile at any 2θ angle where a particular specimen may have profiles that require fitting.

7.8 Modelling the specimen broadening contribution to an X-ray diffraction profile

It has been generally accepted that $f(x)$ may be represented by a Lorentzian function when profile broadening is caused by small crystallite size. When microstrain is responsible, $f(x)$ has often been *assumed* to be represented by a Gaussian function. To test these two profile shape functions, diffraction patterns were obtained by Howard and Snyder (1985b) from three materials. Two commercial aluminas were used: Linde A and C with a nominal crystallite size of 0.3 μm and 1.0 μm respectively. A sample of tungsten with a nearly mono-particle size distribution of approximately 5 μm was also examined. The tungsten sample exhibited a high degree of line broadening due to microstrain and possibly stacking faults. The analysis progressed in the following sequence.

7.8.1 Generation of the instrument function, $g(x)$

Since the $g(x)$ model includes the $K_{\alpha 2}$ and $K_{\alpha 3,4}$ lines, only estimates for the $K_{\alpha 1}$ line positions need be obtained from the pattern. These positions were used to generate profiles from the instrumental function, which is expressed as the four polynomials in 2θ shown in Fig. 7.7, one for each of the split-Pearson variables; $g(x)$ for each line was generated by using one split-Pearson VII function for each wavelength component.

7.8.2 Generation of the specimen profile, $f(x)$

Lorentzian and Gaussian profile functions were generated from their equations, normalized to obtain unit integrated area.

7.8.3 Numerical convolution of $g(x)$ and $f(x)$

In the absence of an analytical convolution function for the profile models, numerical techniques were employed. The $g(x)$ profiles were generated at discrete values, i.e. at the same values of 2θ at which the observed pattern was measured, using eqn (7.12). Since $f(x)$ was taken to be symmetric, j values were generated from $-n$ to $+n$, where n is the number of points in the tails. Profile $f(x)$ was initially also generated at intervals corresponding to that of the pattern; however, this constraint has been removed in later versions of SHADOW. The convoluted profile is obtained from:

$$g * f_{(i)} = \sum_{j=-n}^{+n} g_{(i-j)} f_{(j)}. \tag{7.12}$$

The convolution gathers intensity contributions from all points in both g

and f. The only approximation in the numerical convolution comes from the fact that the n limits are finite. The broader the full width of $f(x)$, the greater is the smearing of the profiles and loss of apparent resolution.

7.8.4 *Profile analysis*

Patterns of the tungsten and alumina samples were collected on an instrument previously calibrated for $g(x)$ using the NIST Si. Each pattern was step-scanned from 20 to 140° 2θ with an angular increment of 0.05° and count time of 20 s per step.

The lines in each pattern were fit using the two models for the specimen profile shape function $f(x)$; $2\theta_K$ and I_0 were refined for the $g(x)$ component, and the Γ for $f(x)$. The least-squares error criterion was the R_{wp} defined in Table 1.3 of Chapter 1.

7.8.5 *Broadening as a function of angle: crystallite size and microstrain contributions*

Since all non-specimen related broadening terms, like the $\tan \theta$ spectral broadening, have been accounted for in $g(x)$, the angular dependence of the pure crystallite size and microstrain effects may be modelled by eqns (7.2) and (7.3). The slope of the curves Γ vs $\lambda/\cos \theta_K$ and Γ vs $k \tan \theta_K$ give $1/\tau$ and ε respectively or, more simply, a Williamson–Hall (1953) plot of $\Gamma_K \cos \theta_K$ vs $k\varepsilon \sin \theta_K$ gives a line of slope ε and y-intercept λ/τ. The results of such a regression analysis indicated that both alumina samples exhibited pure crystallite size broadening while the tungsten exhibited pure microstrain broadening, consistent with our understanding of the samples. Figure 7.8 shows the crystallite size broadened $f(x)$ profile and the final composite (convolution product) envelope matching the observed data for the two alumina samples.

It is interesting to note that the crystallite-size broadened $f(x)$ profile was correctly modelled by a Lorentzian profile. However, rather than the expected Gaussian $f(x)$ profile for the tungsten sample, a Lorentzian was required to fit the data. Even though the functional dependence of the Γ values for the tungsten sample followed the $\tan \theta$ dependence, the Lorentzian nature of $f(x)$ may indicate a more complex origin of the broadening. See Chapter 8 for further discussion of this point and Chapter 11 for a similar finding.

Program SHADOW uses these simple models to allow for the simultaneous determination and refinement of τ and/or ε. The extra degrees of freedom from the addition of these two parameters is more than offset by elimination of the Γ parameter for each line being refined. SHADOW was used to refine those segments over which the profile data were taken. The results of refinement of ε and τ from the non-linear least squares are shown in Table 7.1

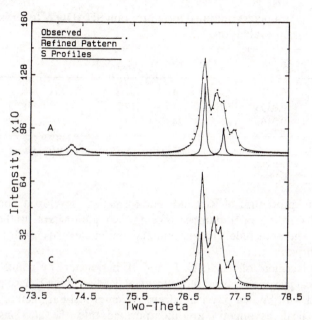

Fig. 7.8 Both Linde A and C aluminas show significant line broadening. The inner curves are the deconvoluted S functions. The integral breadth of these S values show that the X-ray crystallite size of Linde A is about half that of Linde C.

Table 7.1 Parameters obtained from least-squares from crystallite size and microstrain evaluation

Specimen	ε	R (%)	τ (nm)	R (%)	ε	τ (nm)	R (%)
Tungsten	0.0794	5.9	35.5	20.5	0.0793	—	5.9
Linde A	0.0270	44.2	130.9	26.9	—	129.5	26.9
Linde C	0.0144	43.4	244.7	25.3	—	243.4	25.3

Values on the left are for individual refinement of τ and ε, and on the right are the results of simultaneous refinement.

and SHADOW refinements are summarized in Table 7.2. 'Individual' refinement refers to the use of a single broadening function, while 'simultaneous' indicates that both ε and τ were refined.

7.8.6 *Rietveld analysis*

With the $g(x)$ and $f(x)$ functions properly modelled for X-ray diffraction profiles, this method can now be applied to the Pattern-Fitting-Structure-

Table 7.2 Parameters obtained from program SHADOW for crystallite size and microstrain evaluation

Specimen	ε	R_{wp} (%)	τ (nm)	R_{wp} (%)	ε	τ (nm)	R_{wp} (%)
Tungsten	0.0763	7.8	41.0	11.4	0.0763	—	7.8
Linde A	0.0353	17.7	116.5	17.7	—	117.3	18.8
Linde C	0.0169	17.6	248.9	17.6	—	219.9	17.8

Values on the left are for individual refinement of τ and ε, and on the right are the results of simultaneous refinement.

Refinement procedure of Rietveld. The formalism developed above allows for the discarding of a number of empirical parameters in the Rietveld method. These include the asymmetry parameter (A) and the profile half-width parameters U, V, W.

The replacement of the A, U, V, and W parameters by τ and ε removes all empirical parameters from the refinement, leaving in their place parameters associated with known physical effects. All empirical parameters associated with the peak asymmetry are incorporated into the $g(x)$ instrumental function and are fixed during refinement.

When both microstrain and crystallite size effects are present, the integral breadth, β, of the $f(x)$ profile was assumed to be a linear addition of the two components. That is, $\beta_f = \beta_\varepsilon + \beta_\tau$. The basis for this assumption is that the convolution of two Lorentzian functions yields another Lorentzian function. Thus,

$$L(\text{microstrain}) * L(\text{size}) * g(x) = L(\text{combined}) * g(x). \qquad (7.13)$$

The value for the full width of the new convoluted profile is simply the sum of the component β's.

The X-ray version of the Rietveld procedure as developed by Young *et al.* (1977) and Wiles and Young (1981) was modified by Howard and Snyder (1985a,b) to incorporate the $g(x) * f(x)$ convolution and the modelling of $f(x)$ described above. The existing program architecture was preserved so that the convolution procedures can be 'switched' on and off. The following four options are allowed when convolution is 'on':

- no broadening present, i.e. use the $g(x)$ profiles alone,

- the broadening of $f(x)$ is considered as being due to crystallite size,

- $f(x)$ broadening is due to microstrain effects, and

- broadening arises from both crystallite size and microstrain.

Turning 'off' the instrumental convolution allows the program to function as originally designed.

To generate the derivatives necessary for the optimization algorithm, some numerical differentiation was required. The convolution process added to this program was the same as previously employed in program SHADOW. Test refinements were carried out on each of the specimens by varying the same structure parameters. However, the first refinement was performed in the original program context, i.e. using A, U, V, and W. The second refinement used the instrumental function convolution process.

7.8.7 *Discussion of results*

For each of the samples, the broadening of the specimen profile β_f was determined by using both a Gaussian and a Lorentzian specimen function convoluted with the $g(x)$ function. To determine the quality of fit obtained from the refinement, the residual errors were compared, and the differences between refined and observed profiles were visually inspected. Figure 7.8 shows the fitting of selected lines for two of the samples.

In all cases, the use of the Lorentzian function to model $f(x)$ yielded a lower value for the residual errors. Visual comparison showed the Gaussian function to produce distorted representations of the observed lines. For these reasons, further use of the Gaussian function was not considered. Brown and Edmonds (1980) found that for Guinier data, without deconvoluting f, the broadening arising from reported 'grinding' of the material, and therefore presumed to be due to microstrain, was Gaussian, and crystallite size broadening was Lorentzian. In our samples, both broadening terms are Lorentzian in nature. In Chapter 11, David and Jorgensen describe a similar finding even though completely different specimens and instruments were used.

In the Rietveld refinements of the alumina samples, the angular dependence of the broadening could be modelled by the size and microstrain functions (7.2) and (7.3). The alumina samples exhibited almost pure size-determined broadening. Linde A exhibited a higher degree of broadening than that of C. Linde C showed a perceptible degree of broadening over the Si (SRM640a) profiles used to determine $g(x)$.

The broadening of the tungsten profiles appeared to be entirely a result of a microstrain in the sample. This is supported by SEM micrographs and the automatic elimination of crystallite size broadening by the refinement algorithms. Finally, the introduction of an analytical expression describing $f(x)$ broadening simplifies profile refinement. This constrained broadening reduces the number of parameters undergoing refinement while characterizing and quantifying the source of the broadening. However, the cost is a considerable increase in the execution time of the program. To eliminate the

Table 7.3   Results from the Rietveld Structure Refinement with (left) and without (right) $g(x)$ deconvolution

Specimen	ε	τ (nm)	R (%)	ε	τ (nm)	R (%)
Tungsten	0.0315	NA	9.2	NA	NA	7.7
Linde A	NA	134.9	15.4	NA	NA	18.7
Linde C	NA	257.8	15.4	NA	NA	19.6

need for numerical convolution the attractive Tomandl function needs to be incorporated into the Rietveld formalism.

Table 7.3 lists the results from the six refinements performed. As verified by the data in the table, the inclusion of the $g(x)$ deconvolution has lowered the residual errors after refinement. Crystallite size results for the Linde A and C specimens showed a larger degree of broadening in A than in C. A segment of the refined patterns are shown in Fig. 7.8. However, attempts to refine both size and microstrain failed. Since no constraints were available in this version of the Rietveld program, there was a tendency to make the crystallite microstrain parameter (ε) negative. The accompanied lowering of the residual error was done at the expense of introducing a value for microstrain that was physically unrealistic.

The fit of the tungsten pattern again indicated that $g(x)$ deconvolution lowered residual error compared to the other method of refinement. Again, trouble was encountered refining both crystallite size and microstrain parameters. In this case, an unrealistically large value for the crystallite size hindered refinement and was therefore removed from the processing.

In conclusion, the application of instrumental function convolution in the Rietveld structure refinement, aids the description of the profile intensity distributions. Replacement of the numerical parameters A, U, V, and W with analytical parameters characterizing the specimen crystallite size and microstrain was effective in describing the broadening of the specimen profiles as a function of angle.

References

Brown, A. and Edmonds, J. W. (1980). *Adv. X-ray Anal.*, **23**, 361.
Caglioti, G., Paoletti, A., and Ricci, F. P. (1958). *Nucl. Instrum. Methods*, **3**, 223–6.
Cox, D. E., Hastings, J. B., Thomlinson, W., and Prewitt, C. T. (1983). *Nucl. Instrum. Methods*, **208**, 273–8.
Cox, D. E., Toby, B. H., and Eddy, M. M. (1988). *Austral. J. Phys.*, **41**, 117–31.
Delhez, R., de Keijser, T. H., and Mittemeijer, E. J. (1982). *Fres. Z. Anal. Chem.*, **312**, 1–16.

Delhez, R., de Keijser, Th. H., Mittemeijer, E. J., and Langford, J. I. (1988). *Austral. J. Phys.*, **41**, 213–27.

Edwards, H. J. and Langford, J. I. (1971). *J. Appl. Crystallogr.*, **4**, 43–50.

Frevel, L. K. (1987). *Powder Diffract.*, **2[4]**, 237–41.

Hastings, J. B., Thomlinson, W., and Cox, D. E. (1984). *J. Appl. Crystallogr.*, **17**, 85–95.

Hepp, A. and Baerlocher, Ch. (1988). *Austral. J. Phys.*, **41**, 229–36.

Howard, S. A. and Snyder, R. L. (1983). *Adv. X-ray Anal.*, **26**, 73–81.

Howard, S. A. and Snyder, R. L. (1985a). In *Advances in material characterization II* (ed. R. L. Snyder, R. A. Condrate, and P. F. Johnson) pp. 43–56. Plenum Press, New York.

Howard, S. A. and Snyder, R. L. (1985b). NYS College of Ceramics Technical Publication.

Howard, S. A. and Snyder, R. L. (1989). *J. Appl. Crystallogr.*, **22**, 238–43.

Jones, F. W. (1938). *Proc. R. Soc. Lond. Ser. A*, **166**, 16–43.

Khattak, C. P. and Cox, D. E. (1977). *J. Appl. Crystallogr.*, **10**, 405–11.

Klug, H. P. and Alexander, L. E. (1974). *X-ray diffraction procedures* (2nd edn). Wiley, New York.

Ladell, J., Zagofsky, A., and Pearlman, S. (1970). *J. Appl. Crystallogr.*, **8**, 499–506.

Langford, J. I. (1978). *J. Appl. Crystallogr.*, **11**, 10–14.

Langford, J. I. and Louër, J. I. (1991). *J. Appl. Crystallogr.*, **24**, 149–55.

Langford, J. I., Delhez, R., de Keijser, Th. H., and Mittemeijer, E. J. (1988). *Austral. J. Phys.*, **41**, 173–87.

Mallory, C. L. and Snyder, R. L. (1979). *Adv. X-ray Anal.*, **22**, 121–31.

Malmros, G. and Thomas, J. O. (1977). *J. Appl. Crystallogr.*, **10**, 7–11.

Parrish, W., Huang, T. C., and Ayers, G. L. (1976). *Am. Crystallogr. Assoc. Monograph*, **12**, 55–73.

Parrish, W., Hart, M., and Huang, T. C. (1986). *J. Appl. Crystallogr.*, **20**, 79–83.

Pyrros, N. P. and Hubbard, C. R. (1982). *J. Appl. Crystallogr.*, **16**, 289–94.

Rietveld, H. M. (1969). *J. Appl. Crystallogr.*, **2**, 65–71.

Sabine, T. M. (1977). *J. Appl. Crystallogr.*, **10**, 277–80.

Sabine, T. M. (1989). Oral presentation at the International Workshop on the Rietveld Method, June 1989, Petten, The Netherlands.

Scherrer, P. (1918). *Nachr. Ges. Wiss. Göttingen*, 98–100.

Smith, G. S., Johnson, Q. C., Cox, D. E., Snyder, R. L., Smith, D. K., and Zalkin, A. (1987). *Adv. X-ray Anal.*, **30**, 383–8.

Snyder, R. L. (1983). In *Advances in X-ray materials characterization II*, pp. 449–64. Plenum Press, New York.

Taupin, D. (1973). *J. Appl. Crystallogr.*, **6**, 266–73.

Tomandl, R. L. (1987). University of Erlangen, Private Communication.

Wertheim, G. K., Butler, M. A., West, K. W., and Buchanan, D. N. E. (1974). *Rev. Sci. Instrum.*, **45**, 1369–71.

Wiles, D. B. and Young, R. A. (1981). *J. Appl. Crystallogr.*, **14**, 149–51.

Williamson, G. K. and Hall, W. H. (1953). *Acta Metall.*, **1**, 22–31.

Wilson, A. J. C. (1963). *Mathematical theory of X-ray powder diffractometry* (Chapter 4). Philips Technical Library, N. V. Philips, Gloeilampenfabrieken, Eindhoven, 128 pp.

Young, R. A. and Wiles, D. B. (1982). *J. Appl. Crystallogr.*, **15**, 430–8.

Young, R. A., Mackie, P. E., and von Dreele, R. B. (1977). *J. Appl. Crystallogr.*, **10**, 262–9.

8

Crystal imperfection broadening and peak shape in the Rietveld method

Robert Delhez, Thomas H. de Keijser, J. Ian Langford, Daniel Louër, Eric J. Mittemeijer, and Eduard J. Sonneveld

8.1 Introduction

The Rietveld method requires a two-part starting model (see Chapter 1 and elsewhere), a structural model based on approximate atomic positions and a non-structural model which takes into account the contributions of individual line profiles in terms of analytical or other differentiable functions. Both must be considered in order to achieve an optimum representation of the observed pattern. The total intensity of Bragg reflections and, to a first approximation, their positions, are determined by the structural model, but the non-structural model, and hence the representation of diffraction lines, depends on the instrument used and on the microstructural and other properties of the sample. As is also explicitly noted in Chapters 7 and 9, structural imperfections should therefore be taken into account when considering the physical origin of line shapes. These imperfections can be diverse and include: the dimensions and morphology of coherently diffracting domains (crystallite-size effects), variation in interatomic distances due to internal stresses or non-stoichiometry, micro-twinning, stacking faults, dislocations, and other forms of atomic disorder. Microstructural features directly influence the shape of line profiles. The influence depends on the direction and magnitude of the diffraction vector. Therefore, line shape parameters do not vary smoothly with 2θ or $d*$ $(= 2 \sin \theta/\lambda)$.[†] A plot of some line-shape parameter vs 2θ or $d*$ may have an 'irregular' appearance. In what follows, the diffraction effects of structural imperfections are discussed, existing simple approaches to modelling line breadths and line shapes in the Rietveld method are reviewed, and an alternative two-stage approach is presented. This approach starts with a preliminary study of line-profile parameters, by means of pattern decomposition, to ascertain their $\overline{hkl}$ dependence. If it is evident that structural imperfections have a measurable effect on the shape of individual reflections, relationships obtained from pattern decomposition can be used to model the $\overline{hkl}$ dependence of line-profile parameters (breadth and shape) in the Rietveld method. By '$\overline{hkl}$ dependence', a dependence on the direction as well as the magnitude of $d*$ is meant.

8.2 Origins of line-profile shapes and breadths

It is well known (e.g. Wilson 1963; Warren 1969) that the observed pattern ($h(x)$ profiles) is the convolution of instrumental aberration functions ($g(x)$ profiles), including the wavelength distribution and sample-induced geometric and physical effects, with profiles which are specific to the microstructure of the sample ($f(x)$ profiles).

8.2.1 $g(x)$ profiles

The breadth and shape parameters for line profiles due to instrumental aberrations (the instrument-resolution function) vary smoothly with 2θ or $d*$ and can readily be modelled by using data from a suitable standard. This has been treated at some length in the preceding chapter and in Chapter 9 from somewhat different starting points. For a conventional angle-dispersive diffractometer, the variation is typically as shown in Fig. 8.1. The breadth increases with angle over the range shown and at high angles is dominated by spectral dispersion. There would also be an increase at very low angles due to axial divergence (Louër and Langford 1988). In Fig. 8.2 the variation of shape parameters for various functions reflects the Gaussian character of $g(x)$ profiles at low angles, where geometric aberrations are dominant, and the almost Lorentzian (Cauchy) contribution from the wavelength distribution at high angles. Similar curves can be obtained for other sources of radiation and geometries (e.g. Cox et al. 1988).

[†] For angle-dispersive instruments it is customary to consider the variation of line-profile parameters with diffraction angle, 2θ. However, for energy-dispersive data (e.g. from time-of-flight neutron diffractometers) a distance parameter in reciprocal space, $d*$ $(= 2 \sin \theta/\lambda)$, is used. Although the latter applies equally to the angle-dispersive case, both variables will be used throughout this chapter.

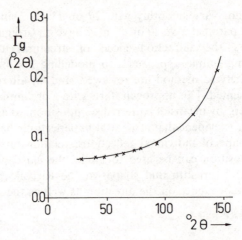

Fig. 8.1 Instrument-resolution function for a conventional diffractometer with an incident-beam monochromator tuned to Cu $K_{\alpha 1}$ radiation, obtained with a BaF_2 standard specimen. ×: observed width; ——: eqn (8.6). (From Louër and Langford 1988.)

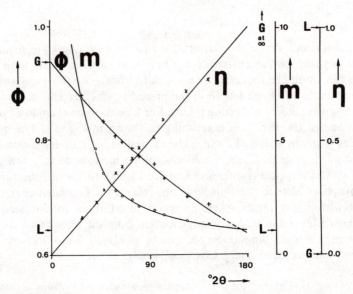

Fig. 8.2 Variation of line-profile shape parameters with 2θ for the same system as in Fig. 8.1. m: Pearson VII index; η: pseudo-Voigt mixing parameter; ϕ: Voigt parameter ($= \Gamma/\beta$); L and G indicate the values of ϕ, m, and η for a pure Lorentzian and a pure Gaussian profile respectively. (See Louër and Langford 1988.)

8.2.2 $f(x)$ profiles

Line broadening due to the microstructure of the sample material in general contains some contributions which are independent of the order of reflection and others which are order dependent. The former arises from the distance, in a direction parallel to the scattering vector, over which diffraction is coherent and can be due to one or more of the following: some weighted average of the actual thickness of the crystallites in this direction, the mean distance between various types of 'mistake', and the reciprocal of the dislocation density. The order-dependent part, on the other hand, represents a variation in d-spacing in the direction considered. This short-range distortion of the separation of planes of atoms can be an actual microstrain, arising from an internal stress distribution, frequently due to the presence of dislocations, or an apparent strain caused by a variation in composition (non-stoichiometry or composition gradient, Mittemeijer and Delhez 1980). For brevity and following customary practice, order-independent contributions will be denoted by 'size broadening' and order-dependent contributions by 'strain broadening'.

8.2.2.1 *'Size' broadening*   The apparent domain size, $\langle D \rangle^{\dagger}$, determined from broadened diffraction lines, is in fact some weighted mean of the length, D, of columns of unit cells in the direction parallel to the diffraction vector, averaged within domains and throughout the sample. In line-broadening analysis one usually distinguishes between the volume-weighted column length, $\langle D \rangle_{v}$, and the area-weighted column length, $\langle D \rangle_{a}$, both being characteristic of the column-length distribution in the sample (see Wilson 1963 and Section 6 of Delhez *et al.* 1980). If the sample contains coherently diffracting domains with a particular shape on average, the apparent sizes $\langle D \rangle_{v}$ and $\langle D \rangle_{a}$ can be expressed as a function of actual geometric parameters which define the morphology of the domain. In general $\langle D \rangle$ is smaller than the actual mean thickness T (cf. Fig. 3 of Langford and Louër 1982). If more than one of the 'size' effects discussed above is present, then the domain size deduced from the breadth of a particular reflection is the harmonic sum and is thus dominated by the smallest dimension. Another factor which can influence the magnitude of $\langle D \rangle$ is multiplicity of reflections (Bertaut 1950): if the domains are other than spherical, the thickness will in general be different for each permutation of the *hkl* reflections with the same d^{*} and again an average is observed. Although domains may have a similar shape, they are likely to have different sizes and $\langle D \rangle$ is thus averaged for all domains within the diffracting volume.

The empirical estimate of 'size' obtained from Γ (=full-width-at-half-

---

† Editor's note: Symbols based on $\langle D \rangle$ are used in this chapter for the various crystallite-size quantities whereas τ has been used uniformly for these quantities in Chapters 7, 9, and 11.

maximum) has no direct physical interpretation, but β (the integral breadth: the width of a rectangle having the same area and height as the line profile) gives the volume weighted average size, $\langle D \rangle_v$. Thus, with β^* as the integral breadth in reciprocal space,[‡]

$$(\beta^*)^{-1} = \langle D \rangle_v = \frac{1}{V} \iiint t \, dx \, dy \, dz \qquad (8.1)$$

where V is the volume of the domain, t is its thickness measured through the point x, y, z in the direction parallel to the diffraction vector, and the integration is over the entire volume of the domain. For example, if there is evidence that on average the domains are approximately spheres of diameter T, then $T = 4\langle D \rangle_v/3$ for all hkl. For other shapes, $\langle D \rangle_v$ will depend on direction and thus on $\overline{hkl}$. For example, the crystallites in a ZnO powder (hexagonal symmetry) could be modelled as cylinders (Langford et al. 1986). The length and direction of the average cylinder derived from the line breadths by a least-square procedure is shown in Fig. 8.3. The curve in this diagram was calculated from the cylinder parameters and shows the dependence of $\langle D \rangle_v$ on lattice direction and thus $\overline{hkl}$; the lengths of the line segments indicate the experimental values of $\langle D \rangle_v$ as obtained from the integral breadths of the reflections indicated.

Since 'size' broadening is strongly influenced by the form of domains, there can be a considerable diversity in line breadth throughout a pattern. The broadening is then said to be 'anisotropic'. (For a general treatment of the theory of 'size' broadening, see Wilson 1962.) As shown above, if the domains are assumed to have a particular shape, this can be modelled on the basis of the dependence of line breadths on lattice direction, transmission electron microscopic studies, and/or the crystal system. The broadening from pure size effects is symmetric, but line profiles broadened by mistakes can be skewed. Line-broadening theory shows that the leading term in a series development for the tails of 'size' profiles is the inverse square of the distance from the peak (Wilson 1962; Vermeulen et al. 1991), and the overall shape tends to be Lorentzian in character (however, see remark on 'size' broadening in Appendix 8.B).

8.2.2.2 *'Strain' broadening*   The profile due to lattice distortion can tend to be Gaussian, but it need not necessarily be symmetric. In general the breadth will not vary smoothly with 2θ or d^*, due to anisotropic elastic properties or non-stoichiometry for example (cf. Section 4.4.1 in Delhez et al. 1980) but will depend on $\overline{hkl}$.

[‡] $\beta^* = \beta \cos\theta/\lambda$, where β is the integral breadth on a 2θ scale and β^* is the integral breadth on a d^* scale; cf. footnote in Section 8.1.

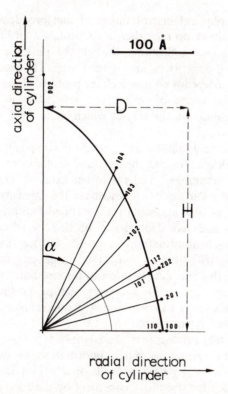

Fig. 8.3   Polar diagram of observed apparent crystallite size $\langle D \rangle_v$ vs angle α between cylinder axis and diffraction vector $\overline{hkl}$ for a ZnO specimen. The curve represents calculated $\langle D \rangle_v$ values for cylindrical crystallites with height $H = 271$ Å and diameter $D = 181$ Å.

In practice either or both categories of sample-imposed broadening may be significant. This can be established by considering the 2θ or d^* dependence of line breadths.

8.3   Line-broadening analysis in conjunction with Rietveld refinement and pattern decomposition

8.3.1   *Introduction*

In the Rietveld method the intensity at each point is calculated from the structural and the non-structural model. Because all possible Bragg reflections are not necessarily observed, particularly in the 'scrambled' high-angle (or large d^*) region of a pattern, some frequently occur only in a mathematical sense. If the observed pattern is precisely described by using both models, then a tentative physical interpretation of line broadening is feasible. In pattern decomposition, on the other hand, atomic coordinates are not used to generate unresolved reflections, though the positions of peaks can be constrained by cell dimensions and space group. The reconstruction of line profiles in this case depends on how well the problem is conditioned. Residual uncertainty in the shape of unresolved reflections is unavoidable and depends fundamentally on the way in which 'scrambled' regions have been reconstructed.

In both the Rietveld method and pattern decomposition the problem of severe overlap is ill-conditioned; the best (and probably only) way to stabilize the treatment of 'scrambled' peaks is to use external information (breadth and shape variation) in order to distribute the intensity of an individual reflection by means of a mathematically defined line profile. How can the angular (or $d*$) and $\overline{hkl}$ dependence of line breadths and shapes be ascertained? Reliable information on profile features can be found from well-defined peaks with little or no overlap. In general, for materials of low crystal symmetry, this only occurs at low or intermediate angles. The Rietveld technique is not appropriate for analysing such peaks, since initially precise atomic coordinates are not known and structure refinement based on a few reflections may well be unreliable.[†]

A more satisfactory procedure is to analyse such peaks by means of pattern decomposition, to extract as much information as possible about the variation of profile parameters with 2θ or $d*$. This leads naturally to a two-stage approach for structure refinement by the Rietveld method. In the first stage line-profile parameters are obtained for as many reflections as can be reliably resolved by pattern decomposition and their angular or $d*$ dependence is studied. If it is evident from these data that the effect of sample imperfections on the distribution of intensity for individual reflections is negligible, then parameters defining the breadth and shape variation can be refined in the usual way in any Rietveld program. If, on the other hand, the effects of microstructural properties of the specimen are found to be significant, the nature of any imperfections present is first ascertained. The relationships of the breadths and shapes of the profiles to hkl are then obtained and used to model these parameters in the Rietveld refinement.

In principle (see below and Chapters 6, 9, and 11), sample imperfections could be modelled directly in the Rietveld refinement, as has been done for a few straightforward cases. However, owing to the potential complexity of defective structures, it would be impracticable to make provision for all possible cases and hence a two-stage approach is recommended.

[†] Editor's note: a detailed analysis of the relative merits of, and relationship between, pattern decomposition and Rietveld analysis is given in Chapter 14.

8.3.2   *Direct approach; incorporation in Rietveld Method*

In the original Rietveld program an approximate model of the idealized crystal structure is refined against the entire diffraction pattern. Both structural and non-structural parameters are refined. The non-structural parameters used in the program constrain the profile breadth to vary smoothly with 2θ or d^*.

In the absence of imperfection broadening, the full-width-at-half-maximum intensity, Γ, for an angle-dispersive instrument, has frequently been modelled as (eqn (1.6))

$$\Gamma = \{U \tan^2 \theta + V \tan \theta + W\}^{1/2} \qquad (8.2)$$

where Γ has been taken on a 2θ scale. This function worked very well with the Gaussian profiles obtained with low or medium resolution neutron diffractometers (Caglioti *et al.* 1958). It is denoted by 'Cag' in Tables 8.1(a) and (b). In the presence of imperfection broadening, such a smooth variation of width with 2θ or d^* can only occur if broadening by the crystal imperfections is 'isotropic' (i.e. independent of lattice direction and hence $\overline{hkl}$). A few procedures based on such imposed isotropy have been developed.

David and Matthewman (1985) adopted a Voigt function for the total broadening ($h(x)$ profiles) and assigned the Lorentzian component and the Gaussian component to 'size' broadening and instrumental broadening respectively, imposing the Scherrer equation (eqns (8.3)) and the Caglioti equation for the dependence of the respective profiles widths on 2θ. A different approach was adopted by Howard and Snyder (1985) who convoluted a Lorentzian $f(x)$ profile, supposed to be due to either 'size' or 'strain', with experimental $g(x)$ profiles, fitted with split-Pearson VII functions, to match the observed $h(x)$ profiles, imposing either the Scherrer equation (for 'size') or the 'tangent' equation (for 'strain'; eqn (8.5)) for the dependence of the width of the Lorentzian $f(x)$ profile on 2θ.

The simultaneous presence of isotropic 'size' and isotropic 'strain' broadening was dealt with by Thompson *et al.* (1987). They used a special pseudo-Voigt function (see Table 1.2) for the total line broadening and assigned the Lorentzian and Gaussian components of Voigt functions 'equivalent to the pseudo-Voigt functions' to 'size' broadening and combined 'strain' and instrumental broadening respectively, imposing dependencies of the widths of these components on 2θ or d^* according to eqns (8.3) and (8.5).

The approaches mentioned above for modelling of isotropic broadening within the Rietveld method are summarized in Table 8.1(a).

In general, materials give rise to anisotropic imperfection broadening and a smooth variation of breadth with 2θ or d^* cannot then be expected.

Table 8.1 (a) Modelling of isotropic size-strain broadening

Instrumental profile		Structural profile		Measured profile	Reference
Shape	2θ Dependence	Shape	2θ Dependence	Shape	
Split P VII	Numerical; fixed	Lorentz (size) or Lorentz (strain)	Sch tan	Numerical convolution	Howard and Snyder (1985)
Gauss	Cag	Lorentz (size)	Sch	Voigt	David and Matthewman (1985)
Gauss	tan	Lorentz (size) and Gauss (strain)	Sch tan	Pseudo-Voigt	Thompson et al. (1987)

(b) Modelling of anisotropic size-strain broadening

Instrumental profile		Structural profile		Measured profile	Reference
Shape	2θ Dependence	Shape	2θ Dependence	Shape	
Lorentz	Cag	Lorentz (size)	Sch; lattice direction dependent	Gauss	Greaves (1985)
Gauss	Cag	Lorentz (size) and Gauss (strain)	Sch; lattice direction independent tan; lattice direction dependent	Pseudo-Voigt	Thompson et al. (1987)
Fourier series	Quasi-Cag	Lorentz (size) and Intermediate Lorentz/Gauss (strain)	Lattice direction dependent Lattice direction dependent	Fourier series	Le Bail (1985)

Ideally, in the Rietveld method the widths of profiles should be refined according to their $\overline{hkl}$ dependence, unlike intensities and, to a first approximation, positions, which are constrained by the crystal structure.

The angular dependence of the profile width due to 'size' broadening is described by the Scherrer equation (denoted by 'Sch' in Tables 8.1(a) and (b)):

$$\text{on } 2\theta \text{ scale:} \quad \text{profile width} = \frac{A'}{\cos\theta} = \frac{A}{\langle D \rangle_v \cos\theta} \tag{8.3a}$$

$$\text{on } d^* \text{ scale:} \quad \text{profile width} = \frac{A'}{\lambda} = \frac{A}{\langle D \rangle_v \lambda} \tag{8.3b}$$

If β or β^* is taken for the profile width, A equals λ. For isotropic 'size' broadening (spherical crystallites/domains) the factor A' is independent of direction in the crystallite/domain. For all other cases $\langle D \rangle_v$ and thus A' depends on $\overline{hkl}$. One of the first attempts to model anisotropic broadening in the Rietveld method was due to Greaves (1985). He assumed diffracting platelets of thickness T and infinitely large lateral dimensions. In this case

$$\langle D \rangle_v = T/\cos\alpha \tag{8.4}$$

where α is the angle between the platelet normal and the diffraction vector. $\langle D \rangle_v$ for cylindrical domains is given by eqns (17) and (18) of Langford and Louër (1982). The integral-breadth apparent size for other regular shapes can be obtained from Langford and Wilson (1978).

In most cases the profile width due to 'strain' broadening is described by

$$\text{on } 2\theta \text{ scale:} \quad \text{profile width} = B' \tan\theta = B\tilde{\varepsilon} \tan\theta \tag{8.5a}$$

$$\text{on } d^* \text{ scale:} \quad \text{profile width} = \tfrac{1}{2} B' d^* = \tfrac{1}{2} B\tilde{\varepsilon} d^* \tag{8.5b}$$

where $\tilde{\varepsilon}$ is some weighted strain (see Section 8.4.2) and B a factor dependent on the profile-width parameter used. This equation is denoted by 'tan' in Tables 8.1(a) and (b). For isotropic 'strain' broadening the factor B' is conceived as independent of lattice direction—and therefore independent of $\overline{hkl}$—and for anisotropic 'strain' broadening it should be taken as $\overline{hkl}$ dependent. In general it is difficult to provide an appropriate $\overline{hkl}$-dependent formulation of B'. For example it has been proposed that a state of stress which is statistically spherical may be adopted. The anisotropy of the elastic constants then leads to $\tilde{\varepsilon}$ and thus B' being $\overline{hkl}$ dependent. Following this approach, Thompson et al. (1987) expressed $\tilde{\varepsilon}$ as a function of $\overline{hkl}$ and refined the parameters based on elastic compliances.

The simultaneous presence of anisotropic 'size' and anisotropic 'strain' broadening was dealt with by Le Bail (1985) and Lartigue *et al.* (1987). 'Size' and 'strain' anisotropies were accounted for by assuming ellipsoids for the average crystallite shape and for the mean square strain. They employed Fourier series to represent the profiles and limited the number of microstructural parameters to be refined by adopting a Lorentzian function for 'size' broadening and an intermediate Lorentz–Gauss function for 'strain' broadening.

The approaches mentioned for modelling of anisotropic broadening within the Rietveld method are summarized in Table 8.1(b) (see also the review by Young and Desai 1989).

Inclusion of some model for imperfection-broadening without *a priori* knowledge of its validity is likely to lead to erroneous results: assuming 'isotropic' 'size' and 'strain' broadening will be unjustified in general and adoption of a specific lattice-direction dependence for 'size' and/or 'strain' broadening can be artificial. A remedy is provided by the two-state procedure discussed below, which is favoured by the present authors.

8.3.3 *Two-stage approach*

The primary reason for modelling crystal-imperfection and instrumental effects in the Rietveld method is to provide an accurate description of line-profile breadths and shapes, and therefore of the associated integrated intensities, and, ideally, of the displacement of peaks from their true Bragg positions. Peak shifts arise from instrumental aberrations, from specimen transparency and displacement, from imperfections (e.g. 'mistakes' and non-stoichiometry) and from homogeneous (macro)strains. Owing to the possible complexity of microstructural features, and the fact that in general the nature of any imperfections present is not known *a priori*, accurate modelling of line profiles for use in a Rietveld analysis can be very difficult to achieve. Therefore a two-stage approach is recommended.

In the first stage, the position, intensity, breadth, and some shape parameter (e.g. Lorentzian/Gaussian fraction or Pearson VII parameter; see Table 1.2) of individual lines are obtained by pattern decomposition (see Chapter 14 and, e.g. Taupin 1973; Parrish *et al.* 1976; Langford *et al.* 1986; Toraya 1986), for which no structural information is required. From these parameters the dependence of breadth and shape on position and direction in reciprocal space (i.e. 2θ or d^* *and* $\overline{hkl}$ dependence) can be determined for all peaks. These parameters can then be predicted for reflections not found during pattern decomposition, owing to too severe overlap or too low intensity. If desired, the results of pattern decomposition can be interpreted in terms of microstructural properties. Also, the pattern can be indexed, if the unit cell is not already known, and precise cell dimensions can be obtained after the line positions have been corrected for systematic errors. The latter may not necessarily be obtained from Rietveld programs with the same accuracy, since refinement of cell dimensions along with other parameters may absorb peak shifts due to any instrumental aberrations for which no allowance has been made, and shifts due to lattice imperfections. If the cell parameters are known, a special, constrained pattern-decomposition method can be applied (Chapter 14 and, e.g. Pawley 1981). Another advantage of using pattern decomposition is that intensities of unambiguously indexed Bragg reflections can be used in *ab initio* structure determination (see Chapter 15). In the second stage, (semi)empirical relationships describing the behaviour of breadths, shapes, and perhaps position of lines, as functions of d^* (or 2θ) and $\overline{hkl}$ are used in the Rietveld refinement.

After the line-profile parameters are obtained by means of pattern decomposition, the variation of breadth and shape parameters with 2θ or d^* is examined. This is an essential preliminary step in all applications of the Rietveld method. By comparing the breadth variation with the resolution curve for the instrument used, the significance of any sample broadening is immediately apparent. Also, from the scatter in the plot (cf. Section 8.4.5), it can be ascertained whether the broadening is 'isotropic' or 'anisotropic'. If practicable, it is clearly desirable at this stage to reduce crystal-imperfection broadening by suitable treatment of the specimen, such as annealing, for example. In any event, the breadth vs 2θ or d^* plot can provide a basis for subsequent modelling of line breadths in Rietveld refinement of the structure.

8.3.3.1 *Isotropic case*

If breadth and shape parameters vary smoothly, with a scatter attributable solely to counting statistics, then microstructural effects are either negligible or their contribution is 'isotropic'. By fitting suitable functions to these curves, an empirical description of the breadth and shape can be incorporated in the Rietveld method. For example the following equations give Γ and η of a pseudo-Voigt function (see Table 1.2) as a function of θ for data obtained with annealed BaF_2, for which sample-imposed broadening was shown to be negligible, and using a conventional X-ray diffractometer equipped with an incident-beam monochromator tuned to Cu $K_{\alpha 1}$ radiation (Louër and Langford 1988):

$$\Gamma^2 = (4.0 \tan^2 \theta + 3.7 + 0.022 \cot^2 \theta) \times 10^{-3} \qquad [\Gamma \text{ in } ° 2\theta]$$

$$\eta \approx 2\theta/\pi \qquad \text{(with } \theta \text{ in radians).} \tag{8.6}$$

The nature of the functions used in the isotropic case is unimportant, provided that residual systematic errors after fitting are negligible.

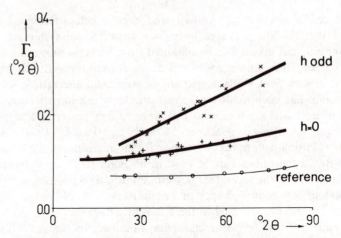

Fig. 8.4 Variation of Γ_h (° 2θ) with 2θ for selected *hkl* diffraction lines of cadmium hydroxide nitrate (from Plévert *et al.* 1989). The reference specimen was an annealed BaF_2 standard.

8.3.3.2 *Anisotropic case*   If the scatter in the Γ vs 2θ or d^* plot is clearly greater than would be expected on statistical grounds, $\overline{hkl}$ (lattice-direction) dependence is indicated. There are then various ways of modelling breadth and shape parameters in the anisotropic case.

1. The simplest, which may be acceptable if the 'anisotropy' is not too severe, is to obtain the average curves and proceed as in the 'isotropic' case. If the widths lie on two or more curves which correlate with particular families of *hkl*, then each can be modelled separately, with provision for the lattice-direction dependence. Such selective line broadening is typical of layer structures with stacking faults. For example (Plévert *et al.* 1989), the broadening for lines with *h* odd from a sample of $Cd_3(OH)_5(NO_3)$ was found to be greater than for *h* zero (Fig. 8.4). (No lines with *h* even could be analysed.) In this case two different functions to describe the two curves shown in Fig. 8.4 were used in the Rietveld refinement and were not allowed to vary. With this two-function model (Fig. 8.5(a)) the weighted pattern *R* factor, R_{wp}, was 7.9 per cent, whereas with a single function (Fig. 8.5(b)) to describe the angular variation of breadths, in which the parameters *U*, *V*, and *W* (see eqn (8.2)) were refined, it was 9.2 per cent. The improvement is clear.

2. For a more rigorous treatment of anisotropic broadening, the breadth variation needs to be examined in detail. Firstly, after correcting for instrumental effects (e.g. Delhez *et al.* 1988) the order-independent and order-dependent contributions to the line profiles are separated, to ascertain whether only one or both are appreciable.

If order-dependent broadening is negligible, the lattice-direction depend-

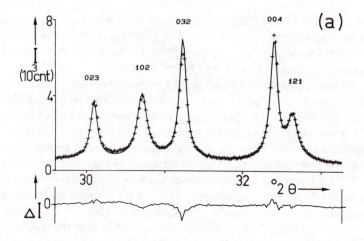

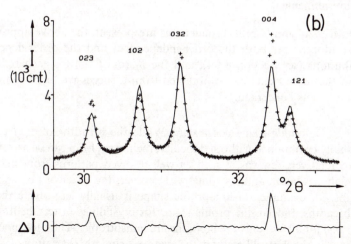

Fig. 8.5 Part of the observed data (I_{obs}: crosses) and the calculated (I_{calc}) and difference ($\Delta I = I_{obs} - I_{calc}$) patterns (upper and lower solid lines) after refinement. The Rietveld method was applied to the same sample as used for Fig. 8.4. Variation of Γ_h represented (a) by a two-function model ($R_{wp} = 7.9\%$) and (b) by a one-function model ($R_{wp} = 9.2\%$).

ence of breadth can be deduced by assuming that on average the domains have a particular shape. From a least-squares comparison of the observed values for domain size and those calculated for the selected shape and orientation of domains with respect to the crystallographic axes, the breadth can be predicted for all reflections (see eqn (8.4) and Fig. 8.3 and eqns 17 and 18 of Langford and Louër 1982) and verified. In the second stage, the value of β_f (or $(\beta^*)_f$) calculated from eqn (8.3) with the 'size' model, can then be 'convoluted' with β_g (or $(\beta^*)_g$), obtained from the instrument-resolution function, to give β_h (or $(\beta^*)_h$) for each reflection used in the structure refinement (see eqns (8.13) and (8.14)).

A similar approach can be adopted for the 'strain only' case, where the assumption of an ellipsoid may suffice to describe the lattice direction dependence of the (mean square) strain component considered. Alternatively, for a given state of stress, an expression for the (mean square) strain component considered can be derived, containing terms depending on $\overline{hkl}$, elastic compliances and stress components. (See also discussion following eqn (8.5).) The number of these terms depends on the crystal symmetry. Then the non-$\overline{hkl}$ dependent parameters in either of the expressions for the 'strain' component can be determined by (least-squares) fitting. In the second stage, the value of β_f (or $(\beta^*)_f$) calculated from eqn (8.5) with the 'strain' model, can then be 'convoluted' with β_g (or $(\beta^*)_g$) obtained from the instrument-resolution function, to give β_h (or $(\beta^*)_h$) for each reflection used in the structure refinement.

If both 'size' and 'strain' broadenings are present, the above approaches can be adopted for both the order-independent and the order-dependent contributions (see Section 8.4.4), and the β_h (or $(\beta^*)_h$) for each reflection used in the refinement can be obtained from a successive 'convolution' of all components concerned.

8.3.3.3   *Line-profile shape parameter*   While the breadths of $f(x)$ profiles can usually be obtained without difficulty, their exact form can be ascertained only in relatively few cases, such as well-resolved patterns from materials with high crystallographic symmetry. However, the effect of non-spherical domains, for example, on line-profile shape is usually less severe than on their breadths. Indeed, the profile shape for crystallites of tetrahedral form is identical for all *hkl*; only the breadths are 'anisotropic'. Thus, for modelling line shape in the Rietveld method, the average ('smooth') variation of a shape factor with 2θ or d^* for the $h(x)$ profiles is probably acceptable in most cases.

8.4   Extraction of 'size' and 'strain' data after Rietveld refinement or pattern decomposition

Note: unless explicitly specified (as in eqns (8.9), (8.10), and (8.25)), β and Γ, and also a (see Sections 8.4.3 and 8.4.4.1), in the equations below denote both crystal-space and reciprocal-space quantities. This implies that the equations and recipes to be presented are valid if these parameters all pertain to crystal space and also if they all pertain to reciprocal space. Recall that d and d^* are related by $d^* = 1/d$.

8.4.1 *Introduction; main conclusion and limitations*

The basis for size-strain analysis in the case in which assumed analytical functions are fitted to (clusters of) diffraction lines, has been outlined in Keijser *et al.* (1983). Two different approaches have been distinguished. These will be dealt with to some extent in Sections 8.4.2–8.4.4. It may be useful to point out at this stage that a procedure suitable in many cases can be summarized as follows (see also Fig. 8.6):

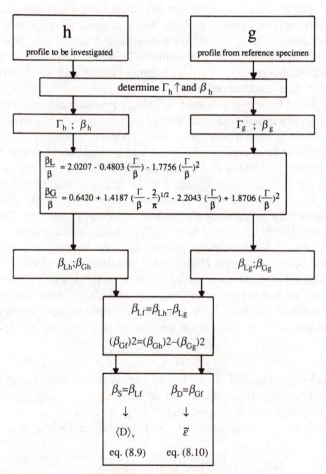

Fig. 8.6 General procedure for the analysis of crystallite-size and microstrain broadening after pattern decomposition applied to data from a specimen showing diffraction-line broadening and from a reference specimen. (Assumed Lorentzian size-broadened profile and Gaussian strain-broadened profile.).

(1) Find the values of Γ ($=$ full-width-at-half-maximum) and β ($=$ integral breadth) for each line to be subjected to line-broadening analysis.
(2) Calculate from Γ and β the auxiliary parameters β_L and β_G for each line.
(3) Correct β_L and β_G for contributions from 'instrumental broadening'.
(4) Calculate domain 'size' and 'strain' parameters from the corrected β_L and β_G.

This procedure is explained in Section 8.4.4.2 and in Fig. 8.6 and an example of its application is discussed in Section 8.4.5.

The adoption of profile-shape functions, as implied by the current versions of the Rietveld method and pattern-decomposition programs, can impose severe constraints on a subsequent size–strain analysis. The rigorous multiple-order methods, such as the Warren–Averbach analysis (see Warren 1969 and reviews by Delhez *et al.* 1980 and 1982), do not impose a profile shape and utilize the difference in order dependence of 'size' and 'strain' broadening to separate 'size' and 'strain' effects.

In the plot of Fourier coefficients vs n, the harmonic number, corresponding to the measured intensity distribution of a single peak, a negative curvature is often observed near $n = 0$ ('hook effect', see e.g. Young *et al.* 1957). This 'hook' effect can have a physical origin (microstrain) but it can also be largely due to instrumental effects and truncation of profile tails. Fitting of a specific profile-shape function can eliminate this negative curvature. It should, however, be realized that the presence or absence of a 'hook' effect in this procedure is fully determined by the mathematical properties of the fitting function employed, no matter how well the function fits and how small or large the effect of the actual truncation. For methods to correct for truncation, see Delhez *et al.* (1986) and Vermeulen *et al.* (1991).

Although large systematic errors in 'size' and 'strain' parameters may result from the assumption of a certain profile shape, it is recognized that such an assumption may have less serious consequences if only relative determinations of 'size' and 'strain' are needed (e.g. for a series of specimens).

8.4.2 *Basis for size–strain analysis*

From the kinematical diffraction theory (e.g. Warren 1969; Wilson 1970) the following two pairs of equations can be derived.

The area-weighted crystallite size, $\langle D \rangle_a$, and the mean square (local) strain, $\langle \varepsilon^2 \rangle$ are given by (eqns (19), (41), and (43) in Delhez *et al.* (1982)):

$$\langle D \rangle_a = -(ld)\left\{ \frac{dA_f(n, l)}{dn}\bigg|_{n\downarrow 0} \right\}^{-1} \tag{8.7}$$

$$\langle \varepsilon^2 \rangle = \frac{-1}{4(\pi l)^2}\left\{ \frac{d^2 A_f(n, l)}{dn^2}\bigg|_{n\downarrow 0} - \frac{d^2 A_S(n)}{dn^2}\bigg|_{n\downarrow 0} \right\} \tag{8.8}$$

where $A_f(n, l)$ and $A_S(n)$ are the normalized cosine Fourier coefficients, evaluated for the interval $l - \frac{1}{2}$ to $l + \frac{1}{2}$, for the composite imperfection-broadened profile f and of the 'size'-broadened component of it respectively, d is the interplanar spacing of the diffracting planes and l is the order of the reflection.

The volume-weighted crystallite size, $\langle D \rangle_v$, and some weighted average strain, $\tilde{\varepsilon}$, can be obtained from (eqns (20) and (29) of Delhez et al. 1982):

$$\langle D \rangle_v = \lambda/(\beta_S \cos\theta) \qquad \text{or} \qquad \langle D \rangle_v = 1/(\beta^*)_S \qquad (8.9)$$

$$\tilde{\varepsilon} = \tfrac{1}{4}\beta_D \cot\theta \qquad \text{or} \qquad \tilde{\varepsilon} = \tfrac{1}{2}(\beta^*)_D/d^* \qquad (8.10)$$

where the subscripts S and D denote the 'size'-broadened and 'strain'-broadened component profiles.

The parameters $\langle D \rangle_a$ and $\langle D \rangle_v$ are different characteristics of the size distribution contained in the diffracting volume; $\langle D \rangle_v$ can easily be 50 per cent larger than $\langle D \rangle_a$ (see Langford and Wilson 1978 and Delhez et al. 1982). The parameters $\langle \varepsilon^2 \rangle$ and $\tilde{\varepsilon}$ are different characteristics of the microstrain distribution contained in the diffracting volume; $\tilde{\varepsilon}$ is 25 per cent larger than $\langle \varepsilon^2 \rangle^{1/2}$ for a Gaussian-strain distribution. (For discussion see Appendix 8.B.) It should be noted that the description of microstrain broadening implied by eqn (8.10) can be ambiguous; see discussion in Section 1 of Langford et al. (1988).

8.4.3   Correction for instrumental broadening

In the measured powder pattern, each line profile, $h(x)$ is the convolution of an instrumental line profile, $g(x)$, with an imperfection-broadened profile, $f(x)$ (see Section 8.2). Normally the $g(x)$ profile is considered as constant for a single $h(x)$ profile. Obviously, for a powder-diffraction pattern covering an appreciable 2θ or d^* range the $g(x)$ profile parameters should be taken as 2θ or d^* dependent.

The $g(x)$ profiles should be recorded from a reference specimen resembling as well as possible the specimen to be analysed but with negligible structural imperfections.

In principle the instrumental line broadening can be removed by a deconvolution operation without additional assumptions (e.g. involving a division of the complex Fourier coefficients of $h(x)$ and $g(x)$ functions: Stokes' (1948) method, e.g. Warren 1969). Within the context of the Rietveld method and pattern decomposition where a profile-shape function is adopted, it appears appropriate to eliminate instrumental line-broadening effects by adopting an analytical profile-shape function for the $g(x)$ profiles as well. Further, it is realized that methods to extract 'size' and 'strain' parameters on the basis of the eqns (8.7) and (8.8) or eqns (8.9) and (8.10) do not require knowledge of the entire $f(x)$ intensity distribution; only derivatives of Fourier coefficients at $n{\downarrow}0$ and β values, respectively, need to be determined. Then, for the majority of the profile-shape functions usually adopted, 'overkill' by full deconvolution can be avoided because an analytical deconvolution can be performed leading to tractable relations among the breadth parameters of the $h(x)$, $g(x)$, and $f(x)$ profiles. Thus, for a Voigt profile-shape function it follows that (Keijser *et al.* 1983)

$$\left.\frac{dA_f}{dn}\right|_{n{\downarrow}0} = -\frac{2}{a}\beta_{Lf} \tag{8.11}$$

$$\left.\frac{d^2A_f}{dn^2}\right|_{n{\downarrow}0} = \frac{4}{a^2}\{\beta_{Lf}\}^2 - \frac{2\pi}{a^2}\{\beta_{Gf}\}^2 \tag{8.12}$$

where (see note at start of Section 8.4) $a = \lambda/(ld\cos\theta)$ in crystal-space, while in reciprocal space a has to be replaced by $a^* = d^*/l$ and

$$\beta_{Lf} = \beta_{Lh} - \beta_{Lg} \tag{8.13}$$

$$\{\beta_{Gf}\}^2 = \{\beta_{Gh}\}^2 - \{\beta_{Gg}\}^2 \tag{8.14}$$

where the subscripts L and G denote the Lorentzian and Gaussian components of the Voigt function. Similar equations hold for the breadths Γ_L and Γ_G. The breadths β_L and β_G of a Voigt function ($h(x)$ or $g(x)$ profile) can be determined from its Γ and β using the following empirical formulae (Keijser *et al.* 1982):

$$\frac{\beta_L}{\beta} = 2.0207 - 0.4803\phi - 1.7756\phi^2 \tag{8.15}$$

$$\frac{\beta_G}{\beta} = 0.6420 + 1.4187\left(\phi - \frac{2}{\pi}\right)^{1/2} - 2.2043\phi + 1.8706\phi^2 \tag{8.16}$$

where $\phi = \Gamma/\beta$.

The analogous procedure for the case of a pseudo-Voigt profile-shape function is given in Keijser *et al.* (1983; eqns (20) and (21)).

8.4.4   *Recipes for size–strain analysis*

As has already been suggested by the treatment in Sections 8.4.2 and 8.4.3, the whole analysis for average 'size' and microstrain values can be based on the breadth parameters Γ and β. As long as some profile shape is assumed for describing each peak in the pattern, the application of potentially rigorous Fourier methods (as the Warren–Averbach analysis; see Warren 1969) in general does not produce better results than breadth methods. As a matter of principle, then, less rigorous, approximate methods are in order.

8.4.4.1   *Approach 1: applying eqns (8.7) and (8.8)*   Adoption of a specific profile-shape function for $h(x)$ and $g(x)$ profiles imples that $A_f(n, l)$ is available in either analytical or numerical form. Hence a value for $\langle D \rangle_a$ is obtained straightforwardly from eqn (8.7).

Clearly, profile-shape functions for which $\mathrm{d}A_f(n, l)/\mathrm{d}n(n \!\downarrow\! 0) \geq 0$ have no physical significance if 'size' broadening occurs. Examples of this category are Gaussian and Pearson VII (with $m > 1$; see Table 1.3) functions. (Further, within the context of the discussion given in the last paragraph of Section 8.4.1, these functions always provide a 'hook' effect!)

A value for $\langle \varepsilon^2 \rangle$ (and $\mathrm{d}^2 A_S(n)/\mathrm{d}n^2$) can be determined from $A_f(n, l)$ data for two orders of a reflection. However, because a profile-shape function *is* adopted in the current Rietveld method and pattern-decomposition programs, here a single-line analysis appears appropriate (see discussion in Section 8.4.1). Furthermore, for many reflections, a second order is unavailable or of low quality. Keijser *et al.* (1983) have proposed the assumption that either

(i)
$$\left. \frac{\mathrm{d}^2 A_S(n)}{\mathrm{d}n^2} \right|_{n \downarrow 0} = 0 \qquad\qquad (8.17)$$

which includes the approximative description of the 'size'-broadened profile by a sum of Laue functions (interference functions of the form $\sin^2(\pi D x)/\sin^2(\pi x)$; see eqn (28) of Delhez *et al.* (1982)), or

(ii)
$$\left. \frac{\mathrm{d}^2 A_S(n)}{\mathrm{d}n^2} \right|_{n \downarrow 0} = \frac{4}{a^2}(\beta_{Lf})^2 \qquad\qquad (8.18)$$

which includes the description of the 'size'-broadened profile by a Lorentz function. (For 'a' see below eqn (8.12).)

Note that with the assumption of eqn (8.17), a Lorentz function for the $f(x)$ profile can never satisfy eqn (8.8) because $\langle \varepsilon^2 \rangle < 0$ is impossible.

The above suffices for any numerical evaluation of Approach 1. Analytical formulae for the cases of pseudo-Voigt and Voigt shape functions are given by Keijser *et al.* (1983).

8.4.4.2   *Approach 2: applying eqns (8.9) and (8.10)*   In this approach it is assumed that the 'size' (order-independent) and 'strain' (order-dependent) broadening of the $f(x)$ profile are described by Lorentzian and Gaussian functions (Keijser *et al.* 1983). (There is some theoretical background for such a supposition; see Appendix 8.B.) Thus

$$\beta_S = \beta_{Lf} \tag{8.19}$$

$$\beta_D = \beta_{Gf}. \tag{8.20}$$

Then, application of eqns (8.9) and (8.10) immediately provides the 'size' and 'strain' estimates $\langle D \rangle_v$ and $\tilde{\varepsilon}$. If the imperfection-broadened profile can be considered as the convolution of 'size' and 'strain' components (cf. discussion in Section 3 of Delhez et al. (1986)), eqns (8.19) and (8.20) in fact imply that the $f(x)$ profile is taken as a Voigt function.

So, if a Voigt function has been adopted for profile-shape description, size–strain analysis can be executed as follows (see also Fig. 8.6):

- determine Γ and β for $h(x)$ and $g(x)$;

- determine β_L and β_G by using eqns (8.15) and (8.16);

- 'deconvolute' by using eqns (8.13) and (8.14);

- determine $\langle D \rangle_v$ and $\tilde{\varepsilon}$ by using eqns (8.19), (8.20), (8.9), and (8.10).

If in the Rietveld method or in the pattern-decomposition method a profile-shape function other than Voigtian is adopted, the analysis on the basis of eqns (8.19) and (8.20) can still be applied, if 'corresponding Voigt profiles' are used. A 'corresponding Voigt profile' is defined as a Voigt profile having the same Γ and β as that of the profile considered. On this basis the following empirical formulae can be derived which yield β_L and β_G of the Voigt profile corresponding to a pseudo-Voigt profile or a Pearson VII profile which have an integral breadth equal to β (Keijser et al. 1983).

For the case of a pseudo-Voigt profile (profile-shape paramter η; see Table 1.2):

$$\frac{\beta_L}{\beta} = 0.017475 + 1.500484\eta - 0.534156\eta^2 \tag{8.21}$$

$$\frac{\beta_G}{\beta} = 0.184446 + 0.812692(1 - 0.998497\eta)^{1/2}$$

$$- 0.659603\eta + 0.44542\eta^2 \tag{8.22}$$

Similar equations for Γ_L and Γ_G have been given by Hastings et al. (1984).

For the case of a Pearson VII profile (profile-shape parameter m; see

Table 1.2)

$$\frac{\beta_L}{\beta} = 0.750445m^{-1} + 0.247681m^{-2} \qquad (8.23)$$

$$\frac{\beta_G}{\beta} = 1.092228 - 1.163332m^{-1} + 1.316944m^{-2} - 1.131115m^{-3}. \quad (8.24)$$

Alternatively, using the ratio Γ/β, the route indicated in Fig. 8.6 could be followed.

So, if a function other than Voigtian has been adopted for the profile-shape description, size–strain analysis can proceed as follows (see also Fig. 8.6):

• determine η or m, or Γ and β, for $h(x)$ and $g(x)$;

• determine β_L and β_G of the 'corresponding Voigt profiles' by using eqns (8.21–8.24), or (8.15) and (8.16);

• 'deconvolute' according to eqns (8.13) and (8.14);

• determine $\langle D \rangle_v$ and $\tilde{\varepsilon}$ by using eqns (8.19), (8.20), (8.9), and (8.10).

8.4.4.3 *Multiple-line method* In the multiple-line approach the contributions to the breadths of f profiles which increase with the distance d^* from the origin of the reciprocal lattice are separated from those which are independent of d^* and then, in some instances, the direction dependence of breadths can be obtained also. The method requires an analytical function (i) which models diffraction-line profiles adequately and (ii) for which the constituent breadths can readily be 'deconvoluted' (cf. eqns (8.13) and (8.14)). A function which in general satisfies both criteria, and which includes the Lorentzian and Gaussian as limiting cases, is the Voigtian (Langford 1978; see also Appendix 8.A). It is found in practice that experimental line profiles often have this form, either precisely, to within experimental error, or to a reasonable degree of approximation. Furthermore, if the $g(x)$ and $h(x)$ profiles are Voigtian, then so is the $f(x)$ profile. The pseudo-Voigt function (Wertheim *et al.* 1974; see also Table 1.2) is a close approximation to a Voigtian (e.g. Keijser *et al.* 1983), but is the summation, rather than convolution, of a Lorentzian and a Gaussian and thus does not comply with the second criterion. The well known technique introduced by Hall for studying the microstructure of metals and alloys, based on β_f or $(\beta^*)_f$ (Williamson and Hall 1953), meets the second of these criteria, but not the first. Hall assumed that the convoluted profiles are all Lorentzian or all Gaussian and in general this approach should only be used for a qualitative, or perhaps semi-quantitative, study of line breadths. Nevertheless, an *hkl*-indexed plot of β_f vs d^* gives a useful overview of the microstructural properties of the sample.

The Voigt model can be used if (Langford 1978)

$$2/\pi \quad \le \phi \ (=\Gamma/\beta) \le 2(\ln 2/\pi)^{1/2} \qquad (8.25)$$

pure Lorentzian   pure Gaussian

and this criterion holds for the majority of diffraction-line profiles encountered in practice. However, as a consequence of the improved instrument resolution and quality of data available nowadays, line profiles for certain samples have been reported for which the intensity in the tails decreases more slowly than a Lorentzian ($\Gamma/\beta < 2/\pi$, e.g. Plévert and Louër 1990), the so-called 'super-Lorentzian' (Wertheim *et al.* 1974). In such cases the best that can be achieved by using existing modelling techniques is to assume that the constituent profiles are Lorentzian. If it has been demonstrated that the Voigtian, or an equivalent function, is an acceptable model, by fitting this function to the $g(x)$ and $h(x)$ profiles by the method of non-linear least squares or a similar process, then the procedure adopted in the multiple-line method is as follows.

(1) The first stage is to obtain β_f and Γ_f from the breadths of the $g(x)$ and $h(x)$ profiles, as in the single-line method. (See Fig. 8.6.)

(2) An *hkl*-indexed plot is then made of $(\beta^*)_f$ vs d^* to obtain a visual overview of the problem. Examples of the interpretation of the Williamson–Hall plot are given by Langford *et al.* 1986.

(3) If $\phi_f \ (=\Gamma_f/\beta_f)$ does not satisfy criteria (8.25) for a particular line profile, the data for this line are reviewed and either remeasured or rejected. The cause is usually 'noisy' data or inadequate fitting of the h profile, though it could of course arise from a non-Voigtian line shape, e.g. a 'super-Lorentzian'.

(4) The Lorentzian (β_{Lf}) and Gaussian (β_{Gf}) components of β_f are obtained from eqns (8.15), (8.16) and (8.13), (8.14).

(5) For Voigtian $f(x)$ profiles, the Lorentzian and Gaussian parts will vary as

$$(\beta^*)_{Lf} = A_L + y_L(d^*)$$

$$\{(\beta^*)_{Gf}\}^2 = A_G^2 + \{y_G(d^*)\}^2$$

where A_L, A_G are constants and y_L, y_G are functions of d^*. The nature of y_L, y_G depends on the sample imperfections which give rise to the d^*-dependent broadening, about which little or nothing is known at this stage. However, if it can be shown that y_L, y_G are linear functions of d^*, then the contributions to β_f can be separated. Plots are thus made of $(\beta^*)_{Lf}$ vs d^* and of $\{(\beta^*)_{Gf}\}^2$ vs $(d^*)^2$, or of the equivalent quantities, if β is in 2θ; in these plots the hkl values are added to each point.

(6) If the plots of (5) are linear for three or more orders of a reflection it holds

$$(\beta^*)_{Lf} = A_L + B_L d^* \tag{8.26a}$$

$$\{(\beta^*)_{Gf}\}^2 = A_G^2 + (B_G d^*)^2. \tag{8.26b}$$

A_L and A_G are the breadths of the Lorentzian and Gaussian components of the Voigtian describing the d^*-independent broadening. B_L and B_G characterize the breadths of the Lorentzian and Gaussian components of the Voigtians describing the d^*-dependent broadening. The A and B values are found from the intercepts and slopes of the plots corresponding to eqn (8.26). If the variation is not linear, then the multiple-line method cannot be used, at least not in the particular form discussed here. In practice it is sometimes found that the variation due to 'anisotropy' of breadth is not large and that the plots are reasonably linear for all hkl or for certain groups of hkl. In such cases the intercepts and slopes yield values which are averaged for all the reflections concerned.

(7) The breadths of the d^*-independent and d^*-dependent components of β_f, $(\beta^*)_{of}$ and $(\beta^*)_{d^*f}$, are obtained from eqn (25) of Langford 1978:

$$p = q \exp(-k^2)/(1 - \mathrm{erf}\, k) \tag{8.27}$$

where $q = A_G$ and $k = A_L \cdot \{\pi^{1/2} A_G\}^{-1}$ for $(\beta^*)_{of}$ $(=p)$. Similarly, $q = B_G$ and $k = B_L \cdot \{\pi^{1/2} B_G\}^{-1}$ for $(\beta^*)_{d^*}$ $(=pd^*)$.

(8) The interpretation of breadth components of the $f(x)$ profiles depends on the nature of the sample and on information obtained in stage (2). $(\beta^*)_{of}$ may be due to the size of the crystallites or domains, when it can be interpreted as a volume-weighted dimension in the direction considered by means of eqns (8.3). (See also Section 8.2.) If sufficient data are available, it may be possible to obtain an indication of the average domain shape, by considering the direction dependence of the value of 'size' for each reflection (e.g. Fig. 8.3). Alternatively, $(\beta^*)_{of}$ may be largely due to the mean distance between 'mistakes' or dislocations, again in the direction considered, from which the probability of the occurence of a fault or the dislocation density can be obtained. $(\beta^*)_{d^*f}$ arises from a variation in d spacing within or between domains. As noted previously, this can be interpreted as a microstrain, through eqns (8.5), or as an apparent strain due to composition variation, depending on the nature and history of the sample. The interpretation is

simplified if $B_L = B_G = 0$, and hence $(\beta^*)_{d*f} = 0$, as frequently occurs for ceramics and other materials which have been formed or produced at high temperatures. In such cases a detailed analysis of 'size' broadening is possible. Another case of interest is $B_L = A_G = 0$; $(\beta^*)_{0f}$ is then purely Lorentzian and $(\beta^*)_{d*f}$ is Gaussian, corresponding to the single-line method discussed in Section 8.4.4.2.

The validity of the method depends only on the precision with which the Voigt function models the $f(x)$ profiles and on the linearity of the $(\beta^*)_{Lf}$ and $\{(\beta^*)_{Gf}\}^2$ plots; no other assumptions are made. Its success, on the other hand, depends critically on the quality of the data and on the ability of the fitting procedure to resolve overlapping peaks reliably. Poor counting statistics can lead to erroneous, and perhaps misleading, results. Nevertheless, if reliable data can be obtained for some 30–40 reflections, including several multiple orders, the multiple-line method is a powerful technique for characterizing imperfections in powder specimens.

8.4.5   Examples

The recommended two-stage procedure for modelling anisotropic broadening in the Rietveld method can be illustrated by considering two samples of ZnO analysed by X-ray diffraction. One pattern shows negligible 'strain' broadening, the imperfection broadening being solely due to 'size' effects. The other pattern exhibits both 'size' and 'strain' broadening. Details of these samples and an analysis of the line breadths are given by Langford *et al.* (1986). (See also Sonneveld *et al.* 1986.)

8.4.5.1   *A case of anisotropic 'size' broadening*   It can be seen from the Γ_h vs 2θ plot (Fig. 8.7) that the broadening due to this ZnO sample is highly 'anisotropic' (i.e. lattice-direction dependent). A Williamson–Hall plot (Langford *et al.* 1986, Fig. 5) suggests that, qualitatively, microstrains are negligible and the crystallites are prismatic, on average, with the prism height greater than its 'diameter'. The simplest approach to modelling the breadth variation in the Rietveld method would be to ignore the anisotropy of breadth and only consider the overall trend, but clearly this is not a good approximation. A better approach is to model the anisotropy by assuming that the crystallites have a particular shape. If they are taken to be cylindrical, on average, the diameter D and height H can be obtained from a least-squares comparison of the observed and calculated values of $\langle D \rangle_v$ for all detectable reflections. The resulting 'fit' for the ZnO data is shown in Fig. 8.3, where $D = 181 \pm 14$ Å and $H = 271 \pm 56$ Å. These values can be inserted in eqns (17) and (18) of Langford and Louër (1982) to obtain $\langle D \rangle_v$, and hence β_f (or $(\beta^*)_f$) according to eqn (8.3), for all reflections required in the Rietveld method. One can then proceed as follows (see Section 8.3.3.2).

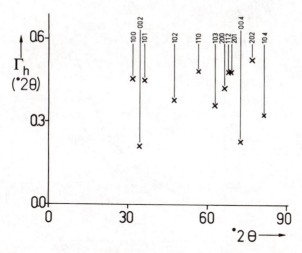

Fig. 8.7 Variation of Γ_h with 2θ for the first 12 reflections for a ZnO sample exhibiting only 'size' broadening (see Section 8.4.5.1), measured with Cu $K_{\alpha1}$ radiation.

In order to find β_h for all reflections in the pattern, it is assumed that all $f(x)$ and $g(x)$ profiles are Voigtian. Values for β_{Lf} and β_{Gf} can be derived from β_f by using eqns (8.15) and (8.16) provided ϕ_f is known. In this case there is little variation in ϕ_f for different hkl and $\langle\phi_f\rangle = 0.67 \pm 0.09$, which is close to the Lorentzian limit ($\phi = 0.64$). The values of β_{Lf} and β_{Gf} can be combined (eqns (8.13) and (8.14)) with the instrumental contributions β_{Lg} and β_{Gg} to give β_{Lh} and β_{Gh}, which are then combined to give β_h by using eqn (8.27). Finally, Γ_h for each reflection can be obtained using eqn (17) (and text following it) of Ahtee et al. (1984). This information is sufficient to construct the $h(x)$ Voigt profile (see Appendix 8.A), if this function is available in the Rietveld program used. If it is not, then some other function with equivalent line-profile parameters will suffice (see Section 8.4.4.2). Most existing programs would require additional code to calculate profile breadths and shapes for each reflection as described above (see also Chapter 14).

8.4.5.2 *A case of anisotropic 'size' and 'strain' broadening* The Williamson–Hall plot for the other ZnO sample (Fig. 6 of Langford et al. 1986) has a positive slope and non-zero intercept, indicating that both microstrains and domain size have a significant effect. Pattern decomposition was carried out by means of the program PROFIT[†] and a Pearson VII function was selected for representing the line shapes. There is little scatter in both curves shown

[†] Copies of the program PROFIT, used to carry out pattern decomposition in this work, can be ordered from E. J. Sonneveld, Laboratory of Metallurgy, Delft University of Technology, Rotterdamseweg 137, 2628 AL Delft, The Netherlands.

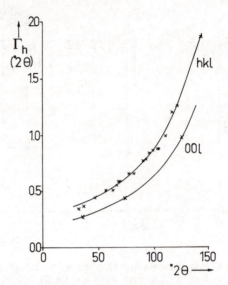

Fig. 8.8 Variation of Γ_h with 2θ for a ZnO sample exhibiting 'size' and 'strain' broadening (see Section 8.4.5.2), measured with Cu $K_{\alpha 1}$ radiation

hkl reflections, except 00l: eqn (8.28)
00l reflections: eqn (8.29).

in Fig. 8.8 for Γ_h vs 2θ, and line breadths can thus be modelled by means of two smoothly varying functions. For *hkl* reflections (except 00*l*), a quadratic in $\tan\theta$ fits the angular variation of $(\Gamma_h)^2$ adequately (Fig. 8.8) and

$$(\Gamma_h)^2 = 0.4632 \tan^2\theta - 0.0065 \tan\theta + 0.1048 \qquad [\Gamma \text{ in } ° 2\theta]. \quad (8.28)$$

Also, for 00*l* reflections,

$$(\Gamma_h)^2 = 0.2248 \tan^2\theta + 0.0222 \tan\theta + 0.0366 \qquad [\Gamma \text{ in } ° 2\theta]. \quad (8.29)$$

Single-line analyses were performed according to Keijser *et al.* (1982) (see Section 8.4.4.2). The pattern-decomposition and size–strain results are listed in Sonneveld *et al.* 1986. The quality of the fit can be assessed from the factors R_p and R_{wp} (Table 1.3; Young *et al.* 1982). The accuracy is lowest at high angles where, as in this case, the intensity of the lines is low and where a considerable peak overlap occurs. This illustrates that for size–strain analysis in conjunction with the Rietveld method or pattern decomposition it is advisable in many cases to apply a single-line analysis only to those lines considered to be fitted adequately. The largest 'size' and smallest microstrain values are observed for the 002 and 004 reflections. In this particular case the other reflections do not display a significant breadth 'anisotropy'.

The refinement of the crystal structure from the powder-diffraction data for this specimen was carried out by means of the Rietveld program DBW3.2S(8804) (an upgraded version of that described by Wiles and Young 1981). The profile–shape function adopted was again a Pearson VII function. The Rietveld refinement involved the following parameters: the z-coordinate of the oxygen atom z_O, two isotropic temperature factors B_O and B_{Zn}, one scale factor, the two cell parameters, one zero-point parameter, one asymmetry factor, six background coefficients and one preferred-orientation factor (in the 00l direction).

If the angular dependence of Γ is as given by eqn (8.28) only, the following R values are obtained:

$$R_p = 5.5 \text{ per cent} \quad \text{and} \quad R_{wp} = 7.8 \text{ per cent.}$$

If, on the other hand, the angular dependence of Γ is as given by eqn (8.29) for the 00l reflection and by eqn (8.28) for the other reflections, then the R values are:

$$R_p = 3.9 \text{ per cent} \quad \text{and} \quad R_{wp} = 5.1 \text{ per cent.}$$

In this case the coefficients of eqn (8.28) were also refined. The final value of the structural parameter z_O is 0.3806(6), B_O and B_{Zn} are 0.63(1) Å^2 and 0.74(1) Å^2, and the values of the crystal-structure indicators R_B and R_F are 2.8 per cent and 1.5 per cent. Figure 8.9 shows graphically the measured data, the fit obtained and the differences between calculated and observed patterns. As expected, a comparison of the R_p values and the R_{wp} values obtained for both refinements shows that incorporating the separate angle dependence of Γ for the 00l reflections improves the agreement between the measured and calculated patterns, in spite of the very small number of 00l reflections.

8.5 Acknowledgement

The authors are indebted to Prof. R. A. Young and Dr P. Desai for the considerable assistance derived from a preprint of their paper (Young and Desai 1989) presented at the 13th Conference on Applied Crystallography, Cieszyn, Poland, 1988.

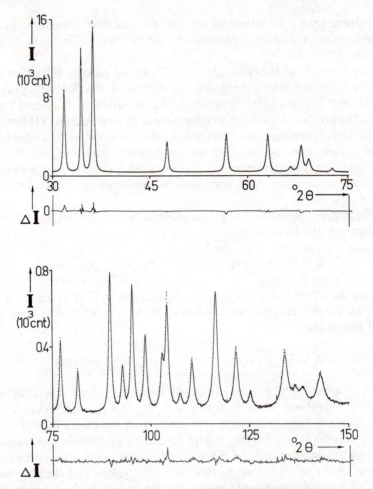

Fig. 8.9 Observed data (I_{obs}: dots) and the calculated (I_{calc}) and difference ($\Delta I = I_{obs} - I_{calc}$) patterns (upper and lower solid lines) after refinement. The Rietveld method was applied to data from a zinc oxide specimen exhibiting 'size' and 'strain' diffraction-line broadening. Both eqns (8.28) and (8.29) were applied for describing the dependence of Γ_h on 2θ. Note the different intensity and 2θ scales for the high-angle range (lower figure).

References

Ahtee, M., Unonius, L., Nurmela, M., and Suortti, P. (1984). *J. Appl. Crystallogr.*, **17**, 352–7.

Bertaut, F. E. (1950). *Acta Crystallogr.*, **3**, 14–18.

Caglioti, G., Paoletti, A., and Ricci, F. P. (1958). *Nucl. Instrum. Methods*, **35**, 223–8.

Cox, D. E., Toby, B. H., and Eddy, M. M. (1988). *Austral. J. Phys.*, **41**, 117–31.

David, W. I. F. and Matthewman, J. C. (1985). *J. Appl. Crystallogr.*, **18**, 461–6.

Delhez, R., Keijser, Th. H. de, and Mittemeijer, E. J. (1980). *Accuracy in powder diffraction*, Natl Bur. Stand. Spec. Publ. 567 (ed. S. Block and C. R. Hubbard), pp. 212–53.

Delhez, R., Keijser, Th. H. de, and Mittemeijer, E. J. (1982). *Fres. Z. Anal. Chem.*, **312**, 1–16.

Delhez, R., Keijser, Th. H. de, Mittemeijer, E. J., and Langford, J. I. (1986). *J. Appl. Crystallogr.*, **19**, 459–66.

Delhez, R., Keijser, Th. H. de, Mittemeijer, E. J., and Langford, J. I. (1988). *Austral. J. Phys.*, **41**, 213–27.

Eastabrook, J. N. and Wilson, A. J. C. (1952). *Proc. Phys. Soc. Lond.*, **B65**, 67–75.

Greaves, C. (1985). *J. Appl. Crystallogr.*, **18**, 48–50.

Hastings, J. B., Thomlinson, W., and Cox, D. E. (1984). *J. Appl. Crystallogr.*, **17**, 85–95.

Howard, S. A. and Snyder, R. L. (1985). *Mater. Sci. Res.*, Symposium on Advances in Materials Research, **19**, 57–71.

Keijser, Th. H. de, Langford, J. I., Mittemeijer, E. J., and Vogels, A. B. P. (1982). *J. Appl. Crystallogr.*, **15**, 308–14.

Keijser, Th. H. de, Mittemeijer, E. J., and Rozendaal, H. C. F. (1983). *J. Appl. Crystallogr.*, **16**, 309–16.

Langford, J. I. (1978). *J. Appl. Crystallogr.*, **11**, 10–14.

Langford, J. I. and Louër, D. (1982). *J. Appl. Crystallogr.*, **15**, 20–6.

Langford, J. I. and Wilson, A. J. C. (1978). *J. Appl. Crystallogr.*, **11**, 102–13.

Langford, J. I., Louër, D., Sonneveld, E. J., and Visser, J. W. (1986). *Powder Diffract.*, **1**, 211–21.

Langford, J. I., Delhez, R., Keijser, Th. H. de, and Mittemeijer, E. J. (1988). *Austral. J. Phys.*, **41**, 173–87.

Lartigue, C., Le Bail, A., and Percheron-Guégan, A. (1987). *J. Less Com. Met.*, **129**, 65–76.

Le Bail, A. (1985). *Proc. 10th Colloque Rayons X*, p. 45. Siemens, Grenoble.

Louër, D. and Langford, J. I. (1988). *J. Appl. Crystallogr.*, **21**, 430–7.

Mittemeijer, E. J. and Delhez, R. (1980). *Accuracy in powder diffraction*, Natl Bur. Stand. Spec. Publ. 567 (ed. S. Block and C. R. Hubbard), pp. 271–314.

Parrish, W., Huang, T. C., and Ayers, G. L. (1976). *Trans. Am. Crystallogr. Assoc.*, **12**, 55–73.

Pawley, G. S. (1981). *J. Appl. Crystallogr.*, **14**, 357–61.

Plévert, J. and Louër, D. (1990). *J. Chim. Phys.*, **87**, 1427–40.

Plévert, J., Louër, M., and Louër, D. (1989). *J. Appl. Crystallogr.*, **22**, 470–5.

Sonneveld, E. J., Delhez, R., Keijser, Th. H. de, Langford, J. I., Mittemeijer, E. J., Visser, J. W., and Louër, D. (1986). *Proc. XIIth Conference on Applied Crystallography*, pp. 26–31. Silesian University in Katowice, Institute of Ferrous Metallurgy, Gliwice, Poland.

Stokes, A. R. (1948). *Proc. Phys. Soc. Lond.*, **61**, 382–91.

Taupin, D. (1973). *J. Appl. Crystallogr.*, **6**, 266–73.

Thompson, P., Cox, D. E., and Hastings, J. B. (1987). *J. Appl. Crystallogr.*, **20**, 79–83.

Thompson, P., Reilly, J. J., and Hastings, J. M. (1987). *J. Less Com. Met.*, **129**, 105–14.

Toraya, H. (1986). *J. Appl. Crystallogr.*, **19**, 440–7.

Vermeulen, A. C., Delhez, R., Keijser, Th. H. de, and Mittemeijer, E. J. (1991). *Proc. 1st Eur. Powder Diffr. Conf.*, 14–16 March 1991, Munich, Germany, Mater. Sci. Forum, **79–82**, 119–24.

Warren, B. E. (1969). *X-ray diffraction*. Addison-Wesley, Reading (Mass.).

Wertheim, G. K., Butler, M. A., West, K. W., and Buchanan, D. N. E. (1974). *Rev. Sci. Instrum.*, **45**, 1369–71.

Wiles, D. B. and Young, R. A. (1981). *J. Appl. Crystallogr.*, **14**, 149–51.

Williamson, G. K. and Hall, W. M. (1954). *Acta Metall.*, **1**, 22–31.

Williamson, G. K. and Smallman, R. E. (1953). *Acta Crystallogr.*, **7**, 574–81.

Wilson, A. J. C. (1962). *X-ray optics* (2nd edn). Methuen, London.

Wilson, A. J. C. (1963). *Mathematical theory of X-ray powder diffractometry*. Centrex, Eindhoven.

Wilson, A. J. C. (1970). *Elements of X-ray crystallography*. Addison-Wesley, Reading (Mass.).

Young, R. A. and Desai, P. (1989). *Arch. Nauk Mater.*, **10**, 71–90.

Young, R. A., Gerdes, R. J., and Wilson, A. J. C. (1967). *Acta Crystallogr.*, **22**, 155–62.

Young, R. A., Prince, E., and Sparks, R. A. (1982). *J. Appl. Crystallogr.*, **15**, 357–8.

Zotov, N. (1991). *Mater. Sci. Forum* **79–82**, 125–30.

8.A   Appendix

The Voigt profile–shape function

The Voigt function, $V(x)$, is the convolution of Lorentzian (Cauchy) and Gaussian functions:

$$V(x) = I_0 \frac{\beta}{\beta_L \beta_G} \int L(z) G(x - z)\, \mathrm{d}z$$

$$= I_0 \frac{\mathrm{Real}\{\exp(-z^2)[1 - \mathrm{erf}\, iz]\}}{\exp(k^2)\, \mathrm{erfc}(k)} = I_0 \frac{\mathrm{Real}\{\omega(z)\}}{\exp(k^2)[1 - \mathrm{erf}\, iz]}$$

$$= I_0 \frac{\beta}{\beta_G} \mathrm{Real}\{\omega(z)\} \tag{8.A.1}$$

with

$$L(x) = \{1 + \pi^2 x^2/\beta_L^2\}^{-1} = \{1 + 4x^2/(\Gamma_L)^2\}^{-1}$$

$$G(x) = \exp\{-\pi x^2/\beta_G^2\} = \exp\{-(4 \ln 2)x^2/(\Gamma_G)^2\},$$

$$z = \frac{\pi^{1/2} x}{\beta_G} + ik$$

$$k = \frac{\beta_L}{\pi^{1/2} \beta_G}$$

β/β_G is given by eqn (8.27) with $p = \beta$ and $q = \beta_G$.

$\omega(z)$ is the complex error function, L and G denote the Lorentzian and Gaussian components of the Voigt function with integral breadths β_L and β_G, and full-widths-at-half-maximum Γ_L and Γ_G. I_0 is the maximum intensity value (at $x = 0$) and β is the integral breadth of $V(x)$. The shape of the Voigt function is usually characterized by $\phi = \Gamma/\beta$ (Langford 1978; see eqns (8.15) and (8.16)).

8.B   Appendix
Shape of the strain-broadened profile

Unnecessary confusion exists in the literature regarding the theoretical background for adopting a Lorentzian (Cauchy) or a Gaussian shape function for the 'strain'-broadened profile. In this appendix we will summarize our present understanding of what is known.

The imperfection-broadened line profile can be regarded as the outcome of the convolution of component line profiles, such as the 'size' and the 'strain' profiles (see Section 8.2). As a consequence, the Fourier coefficients of the total line profile can be represented by the product of the Fourier coefficients of these component line profiles. The most general (i.e. least constrained) theory for combined 'size' and 'strain' broadening is due to Warren and Averbach (cf. Warren 1969).

The distortion cosine Fourier coefficient, $A_D(n, l)$, can be developed as (see eqn (33) in Delhez *et al.* 1982):

$$A_D(n, l) = 1 - 2\pi^2 n^2 l^2 \langle \varepsilon^2(n) \rangle + \tfrac{2}{3}\pi^4 n^4 l^4 \langle \varepsilon^4(n) \rangle + \cdots \qquad (8.B.1)$$

where n = harmonic number, l = order of reflection, and $\varepsilon(n)$ = average strain between two unit cells, n cells apart in the direction of the diffraction vector. Normally (i.e. in the Warren–Averbach analysis), this series is truncated in the following way:

$$A_D(n, l) = 1 - 2\pi^2 n^2 l^2 \langle \varepsilon^2(n) \rangle. \qquad (8.B.2)$$

Hence, eqn (8.B.2) can be used only for small values of nl. Obviously, eqn. (8.B.2) always holds for $n = 0$. It can be shown straightforwardly for $n \geq 1$, that for every $\varepsilon(n)$ for which a Gaussian distribution holds, $\ln A_D(n, l)$ is exactly equal to the last term of the right-hand member of eqn (8.B.2).

At this point the confusion in the literature starts.

As a matter of fact, in the Warren–Averbach treatment *no assumption on the type of $\varepsilon(n)$ distribution is made*, contrary to what many have said. Distribution functions (for $\varepsilon(n)$) other than Gaussian may be nearer to practice (e.g. see Williamson and Smallman 1953). Further, *there is no a priori need for the shape of the distribution function for $\varepsilon(n)$ to be independent of n* (see Delhez *et al.* 1980).

With respect to the last point, it is important to remark that shortly after the Warren–Averbach analysis was presented (1950–2) Eastabrook and Wilson (1952) showed that on statistical grounds the following approximation holds for *larger values of n (and l)*:

$$A_D(n, l) = \exp(-\text{const. } n). \tag{8.B.3}$$

The last equation describes the Fourier transform of a Lorentz function. The above considerations lead to the following summary:

- Equation (8.B.2) is valid for small values of n; it thus *also* holds if the distribution function of $\varepsilon(n)$ *for small values of n is Gaussian*, but it is not necessarily limited to this case.

- Equation (8.B.3) is valid for larger values of n; it thus *also* holds if the distribution function for $\varepsilon(n)$ *for larger values of n is Lorentzian*, but it is not necessarily limited to this case.

The origin of the much used equation that relates a breadth parameter and some parameter describing microstrain (eqn (3.4) and see Chapters 7, 9, and 11) is obscure (see Langford *et al.* 1988). Hence a sound discussion of line broadening due to inhomogeneous strain (microstrain) should not be based on an equation of the form

$$\text{breadth } (2\theta) = \text{const. } \tilde{\varepsilon} \cdot \tan \theta \tag{8.B.4a}$$

or

$$\text{breadth } (d^*) = \text{const. } \tilde{\varepsilon} \cdot d^* \tag{8.B.4b}$$

but such a discussion should rather be based on eqn (8.B.1).

Now, if it is assumed that the distribution function of $\varepsilon(n)$ is Gaussian and *independent of n* (then $\langle \varepsilon^2(n) \rangle = \langle \varepsilon^2 \rangle$ = mean square of the *local* strain), it can be shown that:

$$\tilde{\varepsilon} = \tfrac{1}{2}(2\pi)^{1/2} \langle \varepsilon^2 \rangle^{1/2}. \tag{8.B.5}$$

For this case it also follows that the 'strain'-broadened profile is Gaussian (see Delhez *et al.* 1982 and see text below eqn (8.10) in the present paper). The above paragraph describes the only case known to us where the 'strain profile' is *exactly* Gaussian.

In the original paper on the single-line analysis applying Voigt functions it was stated (de Keijser *et al.* 1982): '... it is assumed that the Cauchy component of the profile is solely due to crystallite size and that the Gaussian contribution arises from microstrain. There is some theoretical justification for this assumption and experimental evidence has also been reported . . .'. There is no reason to abandon this point of view for the reasons that follow.

Let us first discuss 'size' broadening. In a previous paper (Delhez *et al.* 1982) two hypothetical cases of 'size' broadening in the presence of a size *distribution* are considered. It is suggested (see Fig. 9 in that paper) that, particularly for a wide size distribution, a Lorentz function can describe the 'size'-broadened profile reasonably well. However, if a narrow size distribution (the extreme case is a single size/column length) occurs, the Lorentz (Cauchy) function is inadequate. (This is also suggested from simulations presented by Langford and Wilson 1978.) This indicates that the adoption of a Lorentz shape function for 'size' broadening should be accepted or questioned in a way similar to that for the adoption of a Gaussian shape function for strain broadening (see next paragraph).

Now, considering the case of 'strain' broadening, it has been indicated above that, if special conditions are satisfied, a 'Gaussian-strain' profile can occur. But, analogous to the case of the Lorentz 'size'-broadened profile discussed above, it is impossible to state that this would always be the case. (See, in particular, the above discussion of eqns (8.B.2) and (8.B.3).) Then one can turn to experimental data. There is ample evidence that at least in a number of rather different cases, the microstrain-broadened profile can be described reasonably well by a Gaussian function. In the original single-line Voigt paper (de Keijser *et al.* 1982) three totally different experimental cases are presented where an apparent 'Gaussian-strain' profile occurs: microstrain data from several orders of the same reflection coincide (see Tables 2–4 and references given in that paper). A very recent example of 'Gaussian-strain' broadening is provided by Zotov (1991). Of course, it should be realized that other shapes for the 'strain' profile are possible for other specimens. For example, an $h(x)$ profile with a pseudo-Voigt mixing parameter (η) of 0.50 (0.39 for the corresponding $f(x)$ profile) for a sample which exhibited negligible 'size' broadening has been reported by Plévert and Louër (1990).

It is fair to say that a Gaussian-shape function for the 'strain' profile is an assumption subject to the same criticism as the assumption of a Lorentzian function for the 'size' profile. On the other hand, if, in the course of an analysis, the profile shape has to be described by a shape function of the strength/simplicity of the Voigt function, in the light of the existing experimental data, it then appears justified to assume that the 'size' profile *in general* may be described by the 'corresponding' Lorentzian component and that the 'strain' profile *in general* may be described by the 'corresponding' Gaussian component. (For the meaning of 'corresponding' see Section 8.4.4.2.)

Finally we would like to remark that if shapes are assumed for 'size' and 'strain' profiles in the analysis, the data obtained should be checked for internal consistency: e.g. for single-line analysis, 'size' and 'strain' parameters from different orders of reflection should coincide. If this is not the case, better performing shapes have to be adopted. If severe constraints are applied to the shape of line profiles, as in some size–strain analyses and in current versions of Rietveld refinement programs, it is of no use to quarrel about perfect shape functions. The only point to decide is which shape function (of all the ones available) is the most suitable for the specimen concerned.

9

Bragg reflection profile shape in X-ray powder diffraction patterns

P. Suortti

The factors influencing the observed reflection profile shape are divided into two categories: external and internal (intrinsic) or instrument factors, and sample effects (see e.g. Chapters 7 and 8). The instrument factors include the distributions of position, direction, wavelength, and polarization of the incident beam, the transformation of these distributions at the sample due to non-ideal geometry, and recording by the analyser/detector. For a well-characterized instrument, these factors are known, and the external profile shape function can be calculated by ray-tracing. Various methods of calculation are presented.

In this chapter, the sample effects are divided into those affecting the Bragg reflections and those contributing to the background. The incoherent part of the background is calculated from the known incident X-ray flux and composition of the sample, and is subtracted from the measured pattern. The remaining part, Bragg plus thermal diffuse scattering (TDS), is the observed pattern used in the Rietveld refinement. These components can be separated from each other by modelling TDS with a function which describes the fluctuations of scattering from acoustic phonons, and various possibilities are discussed below.

Idealized models are used here to describe the factors that broaden the Bragg reflections (crystallite size, faulting, and strains), and their effects on the profile shapes are modelled by convolutions of Gaussian and Lorentzian functions. Within these approximations, the profile function is Voigtian, where the widths of the Gaussian and Lorentzian parts are given by simple angle-dependent functions involving three parameters.

For use in Rietveld refinement, the above ideas are combined in an approach wherein the incoherent background is subtracted from the observed pattern and the parameters of the line shape function are measures of crystallite size and microstrain. The fact that the TDS is the part missing from the Bragg reflections due to thermal motion provides a constraint, which is used for a self-consistent division of the total pattern into the TDS and the pattern of Bragg reflections (plus non-TDS background).

9.1 Introduction

The Rietveld method and other whole-pattern-fitting schemes are based on a synthetic approach, where the total observed scattering is modelled. Each component, Bragg, thermal diffuse, disorder, inelastic, and resonant scattering has its characteristic dependence on the scattering vector $\mathbf{k}$. The Bragg reflections are 'the signal' and the rest is termed 'the background'. There are different degrees of sophistication in modelling and separation of the background, from simple straight line interpolation to functions following the apparent modulations of the background (Chapter 6). The profile shapes of the Bragg reflections are convolutions of a certain number of functions with assumed angular dependencies, and the parameters of these functions are found by a least squares fit of the model to the pattern of the Bragg reflections. In a second approach, which may be called analytical, the line shapes are decomposed into functions which are calculated from the known diffraction geometry and known properties of the sample (Chapter 7). Ideally, by successive deconvolutions, the intrinsic pattern, which is due to the crystal structure and the crystallite size and microstrain distributions of the sample, would be found, and it would be the 'observed' data for the Rietveld refinement. The third approach is that of 'learned peak shapes' (Hepp and Baerlocher 1988), where the analytic profile shape functions are avoided altogether by interpolating the profile shape function from a few well-resolved reflections of the pattern.

The aim in this chapter is to determine the best compromise between the various approaches. The starting point is the analytical approach; the components of the profile shape are calculated as far as possible. However, the intrinsic pattern is not deconvoluted from the observed pattern with these functions. Actually, the pattern is synthesized from the calculated functions and those describing the effects of crystallite size and microstrain, but only the parameters of the latter functions are varied in the refinement of the model. This approach minimizes the number of parameters to be refined while retaining those which characterize the powder sample. The success of modelling the profile shapes and the physical significance of the model parameters depends critically on an adequate modelling of the background, so this will be discussed at length.

9.2 Instrument function

The contribution of a well-characterized powder diffraction instrument to the observed profile shape, conveniently approximated with various analytical functions (Chapters 7 and 8), can be calculated precisely. The method is called ray-tracing, where the paths of a representative set of X-rays or neutrons are followed from the source to the detector. A graphic variant of the method is phase-space analysis, wherein the contour of the beam in the position–angle–wavelength space is transformed by the optical components. Cartesian coordinates are used: z-axis coincides with the centre of the beam, x is the horizontal and y the vertical coordinate of a ray. The direction of the ray is given by $x' = \mathrm{d}x/\mathrm{d}z$ and $y' = \mathrm{d}y/\mathrm{d}z$. When the beam is reflected by a crystal or mirror, the new coordinates (x, x', y, y') refer to the central ray of the reflected beam. The wavelength of the ray is invariant, and it is correlated to the direction (x', y') through the Bragg law in the case of reflection from a crystal. In the case of a synchrotron source, the position and direction of a ray are correlated, but still in most cases the horizontal and vertical coordinates can be separated. At a given z, 2-dimensional sections in the (x, x', λ) or (y, y', λ) space give adequate descriptions of the beam. The contour diagram of the beam rotates clockwise with z in the (x, x') or (y, y') plane, because for a given ray, $x = x_0 + zx'$ and $y = y_0 + zy'$. In the X-ray case, separate analyses for the two states of polarization should be carried out.

The method is illustrated by an example in which the profile shape is solely determined by the instrument function, namely energy dispersive diffraction. Radiation from a high-voltage W tube is used, so that the energy resolution of the detector can be made comparable to the energy spread due to the beam divergences. The cross-terms between the various components were found to be negligible, so the instrument functions could be calculated separately. The total instrument function was found by successive convolutions. The sample was a pressed pellet set in symmetrical transmission geometry, and the equatorial (horizontal) divergence of the scattered beam was limited by a Soller slit (Honkimäki 1990). The effects of the horizontal divergence in the angle scale are shown in Fig. 9.1(a) and the corresponding broadening due to the vertical divergence in Fig. 9.1(b). In this geometry, the asymmetry of the profile is due to vertical (axial) effects only. A comparison of the calculated profiles with the experimental ones in Fig. 9.2 demonstrates that the instrumental profile shape can be calculated exactly when the diffraction geometry, wavelength distribution, polarization, and response of the detector are known. This is of special importance, of course, when overlapping reflections are resolved, such as 331 from 420 in Fig. 9.2(c).

Phase-space diagrams can be used for precise calculations of the instrumental line shape, if the intensity distribution inside the beam contour is

(a)

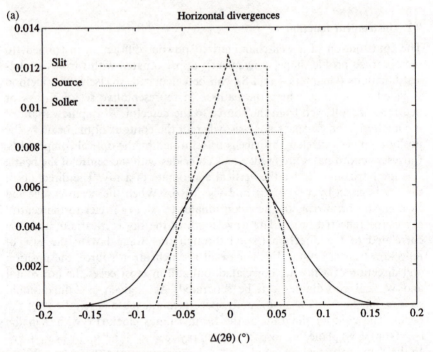

(b)

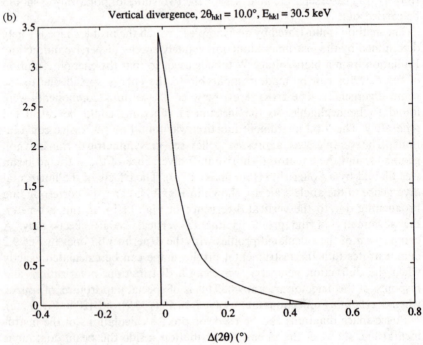

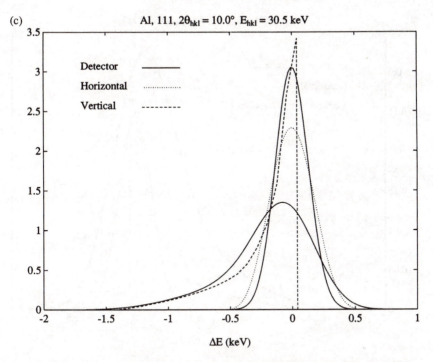

Fig. 9.1 Effects of the beam divergences on the observed energy distribution of X-rays diffracted from a thin powder slab in symmetrical transmission geometry. The scattering angle 2θ is $10°$ and the average energy 30 keV. The apparent source is 1.5×1.5 mm^2, the first slit, at 700 mm from it, had dimensions 0.96 mm (hor.) $\times$ 10.6 mm (vert.), the horizontal opening of the Soller slit was $0.16°$, and the height of the receiving slit, 185 mm away from the sample was 9.0 mm. The variation of the Bragg angle due to horizontal divergence is shown in (a), that due to the vertical divergence in (b), and the convolution of the component functions in energy scale in (c).

uniform or of a simple functional form. In any event, estimates for the width of the instrument function can be obtained in short order. Two examples are shown in Figs 9.3 and 9.4. First, the asymmetric broadening in the Bragg–Brentano geometry due to the flat surface of the sample is shown. The source is an X-ray tube emitting approximately uniformly to half-space, and the beam is limited by a symmetrical horizontal slit. The beam is intercepted by a flat powder sample, and its contour in the phase-space diagram is slightly curved and asymmetric, because the sample surface does not coincide with the Rowland circle. Upon a symmetrical reflection, x changes to $-x$, and the contour of the beam will have rotated in the vertical position at the parafocus. The reflected beam is scanned by a positional slit without any limitation to the divergences of the recorded beam. The instrument function can be calculated from the diagram, as it is proportional

(a)

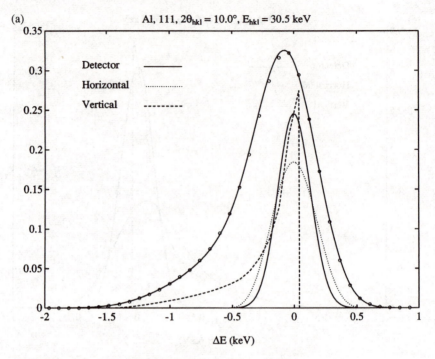

Al, 111, $2\theta_{hkl} = 10.0°$, $E_{hkl} = 30.5$ keV

Detector ——
Horizontal ·········
Vertical - - - -

ΔE (keV)

(b)

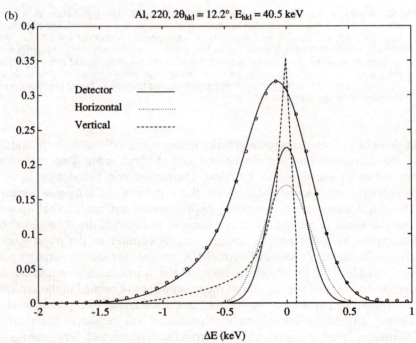

Al, 220, $2\theta_{hkl} = 12.2°$, $E_{hkl} = 40.5$ keV

Detector ——
Horizontal ·········
Vertical - - - -

ΔE (keV)

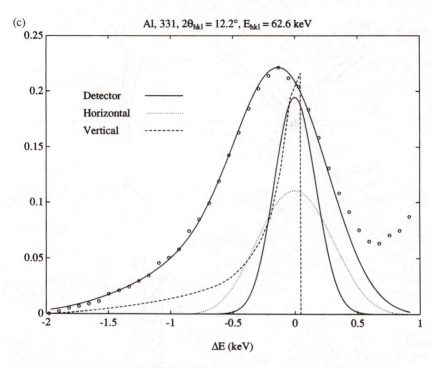

Fig. 9.2 Calculated and observed (circles) profile shapes of a few reflections from an Al powder sample. The overlapping tail of reflection 420 is seen in (c).

to the phase-space area of the beam intercepted by the receiving slit. The effects of the widths of the source and the receiving slit are included in this calculation. Another example is that of using an analyser crystal on the detector arm. The incident beam from a synchrotron source is mono-chromatized by a pair of flat perfect crystals, which have the Bragg angle θ_M. The flat analyser crystal replacing the receiving slit passes only the rays incident at the Bragg angle on the crystal, no matter where on the crystal they fall. The analyser crystal is an 'angle slit', so that the aberrations arising from the finite scattering volume of the sample are eliminated (Hastings *et al.* 1984). The phase-space diagrams are now given in the (y', λ) section (Fig. 9.4), because only the angular coordinate is relevant. There is a wavelength gradient across the beam, due to reflection of a polychromatic beam at the monochromator. At the reflection from a powder sample the angular spread due to the wavelength gradient is either compensated in the $(+, -)$ setting or enhanced in the $(+, +)$ setting. A similar situation occurs at the analyser crystal, which is a 'window' in the (y', λ) plane. When this window is rotated it sweeps through the wavelength range of the reflected beam. It is found

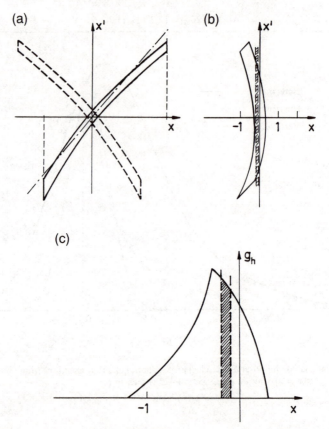

Fig. 9.3 Phase-space diagrams showing the effects of equatorial (horizontal) aberrations on the line shape. The coordinates are relative to the position and direction of the centre ray. The X-ray beam is assumed to be of one wavelength and the intensity distribution uniform. The contour of the beam in the position-angle space (x, x') is given at the sample before (solid line) and after (broken line) reflection in (a), and at the receiving slit in (b); the slit is indicated by broken lines. The intensity distribution recorded by a narrow receiving slit is shown in (c) and a finite slit is indicated by shading.

that the angular width of the scan in the non-dispersive $(+, -, +)$ setting is

$$\Gamma = [\phi_v^2(2 \tan \theta/\tan \theta_M - \tan \theta_A/\tan \theta_M - 1)^2 + \Gamma_{\min}^2]^{1/2}, \quad (9.1)$$

which is the formula given by Cox *et al.* (1988). Here ϕ_v is the vertical divergence of the incident beam, so that $\phi_v/\tan \theta_M = (\Delta E/E)_M$ is the relative energy resolution of the monochromator, when its Darwin width is ignored. The Bragg angle of the powder reflection is θ, and $\Gamma_{\min}$ counts for the effects of the Darwin widths of the monochromator and analyser. The nearly rectangular functions of wavelength spread and the reflectivity curves are

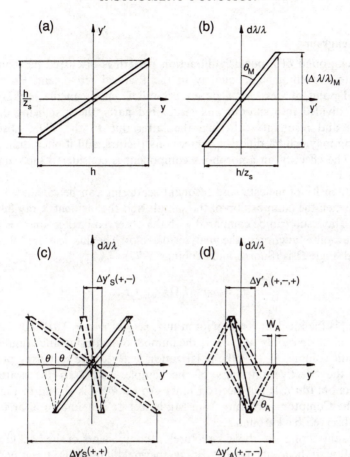

Fig. 9.4 Phase-space diagrams for various combinations of a flat crystal monochromator (Bragg angle θ_M), powder sample (Bragg angle θ), and the crystal analyser (Bragg angle θ_A) in a powder diffraction measurement with synchrotron radiation. The contour of the beam is shown at the sample in (a) and (b) in position-angle (y, y') and angle-wavelength $(y', d\lambda/\lambda)$ sections, respectively. Reflection is shown in (c) indicating the reflected beams (broken lines) in the non-dispersive $(+, -)$ and dispersive $(+, +)$ settings. The beam in the $(+, -)$ setting is shown in (d) with the analyser 'window' of width w_A (broken lines), again in non-dispersive $(+, -, +)$ and dispersive $(+, -, -)$ settings. Here $\Delta y'_s(+, -) = |h/z_s - 2 \tan \theta (\Delta\lambda/\lambda)_M| = (h/z_s)|2 \tan \theta/\tan \theta_M - 1|$ and $\Delta y'_A(+, -, +) = (h/z_s)|2 \tan \theta/\tan \theta_M - 1 - \tan \theta_A/\tan \theta_M|$.

convoluted as Gaussians in eqn (9.1). That is not a bad approximation, as is seen in Fig. 9.1(a), but precise calculations based on Fig. 9.4 are also possible. In general, the phase-space diagrams can be used for optimization of X-ray optical systems as the roles of the separate optical elements are more evident than they are in a straightforward ray-tracing calculation (Suortti and Freund 1989).

9.3 Background

The background of a powder diffraction pattern is discussed here only in regard to modelling it adequately in the Rietveld refinement. From the practical point of view, for a purely crystalline specimen the background may be divided into smooth and structured parts, the first being due to inelastic and resonant scattering, the latter due to the TDS. Disorder scattering may exhibit different degrees of structure, and if often difficult to model. The effect of an amorphous component is considered separately in Chapter 6.

The intensity of inelastic and resonant scattering can be calculated from theory, when the composition of the sample and the incident X-ray flux are known. The result can be compared with the observed background weighted by the response function of the analyser/detector. The flux scattered through the solid angle Ω is (Suortti and Jennings 1977)

$$n_j = n_0 r_0^2 M_0 \Omega K_{\text{pol}, j} t_j I_j, \tag{9.2}$$

where n_0 is the incident X-ray photon flux, $r_0 = e^2/mc^2 = 2.818 \cdot 10^{-5}$ Å the classical electron radius, $M_0 = 1/V_c$ the number of scattering units (unit cells) in the unit volume, $K_{\text{pol}, j}$ is the polarization factor of the scattering process j, t_j is the effective thickness of the sample, and I_j is the scattering cross-section, $(\mathrm{d}\sigma/\mathrm{d}\Omega)_j$, in electron units. An example is shown in Fig. 9.5, where the Compton scattering from Mg is seen to account for a large part of the observed background.

After subtracting the above structureless components of the background, one is left with the total coherent, Bragg, thermal diffuse (TDS), and possible disorder scattering. To a good approximation, the TDS is equal to the part of the Bragg scattering lost due to thermal motion, i.e. the average cross-section is

$$I_{\text{TDS, ave}} = f^2(1 - \mathrm{e}^{-2M}), \tag{9.3}$$

where the case of a monoatomic crystal is taken to make the notation simple; f is the atomic scattering factor and e^{-2M} is the Debye–Waller factor. The TDS arising from acoustic phonons peaks at the Bragg reflections, and its intensity is proportional to the respective Bragg intensity, so that this part of the TDS has the same periodicity as the Bragg scattering. The rest of the TDS is a rather smooth (in 2θ) contribution from optical phonons and multi-phonon processes. Now, an interesting possibility arises: as the shape of the TDS in the vicinity of the Bragg reflections is known, could it be included in the line shape function of the reflection profiles? The powder average

(a)

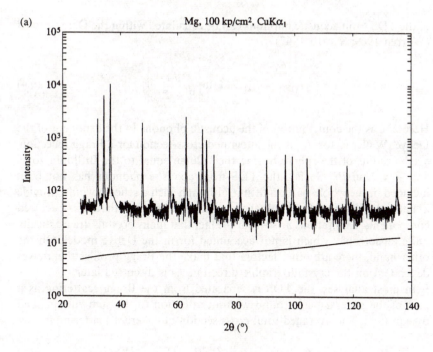

(b)

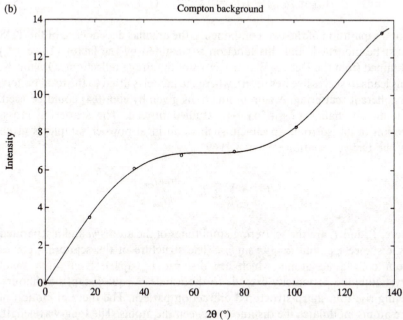

Fig. 9.5 (a) The powder pattern of Mg and (b) the calculated intensity of the Compton scattering.

of the TDS from acoustic phonons can be calculated within the Debye model
(Warren 1969; Suortti 1967)

$$I_{TDS1} = I_{hkl} \cdot 2M_a \ln \left| \frac{g_m}{R - R_{hkl}} \right|. \qquad (9.4)$$

Here $2M_a$ is the contribution of the acoustic phonons to the exponent of the
Debye–Waller factor, I_{hkl} is the integrated cross-section for a Bragg reflection,
g_m the radius of the sphere having the volume equal to the Brillouin zone,
and $R = 2 \sin \theta/\lambda$. So far, the TDS from acoustic phonons has not been
included in the whole-pattern fitting, although such a scheme would provide
a natural division of the observed pattern into smooth and structured
components. Limitations arise from the fact that many crystals are elastically
quite anisotropic, which is not accounted for in the Debye model. On the
other hand, there are other factors that make the Bragg profiles themselves
dependent on the crystallographic direction, as is discussed later.

In most analyses, the TDS is separated from the Bragg scattering as a
part of the general background without attention to the sum rule implied
by eqn (9.3). The averaged total cross-section of coherent scattering is

$$I_{coh} = I_{Bragg} + I_{TDS} = f_e^{2-2M} + (1 + e^{-2M})f^2. \qquad (9.5)$$

For a separation of the two components, the angular dependence of the TDS
must be modelled, and this function is weighted by the factor $(1 - e^{-2M})$
obtained from the Debye–Waller factors of the Bragg reflections. Obviously,
this leads to successive iterations, where the model is fitted to the total pattern
of coherent scattering. A sum of functions given by eqn (9.4) could be used,
but an alternative possibility was studied instead. The scattering cross-
section of an isotropic specimen, such as an ideal powder sample, is given
by the Debye equation (Warren 1969)

$$I_{coh} = \sum_m \sum_n f_m f_n \frac{\sin kr_{mn}}{kr_{mn}}. \qquad (9.6)$$

Here, f_m and f_n are the scattering amplitudes of the atoms m and n separated
by distance r_{mn}, and $k = 4\pi \sin \theta/\lambda$. The structure of the scattering comes
from correlating atoms which are distance r_{mn} apart. Even in a small
crystallite, the periodicity extends to thousands of equi-distant neighbours,
giving rise to a sharp-structured diffraction pattern. The thermal motion of
the atoms modulates the distances between the atoms. The long-wavelength
acoustic waves, which make the neighbouring atoms move in the same
direction, change the distances only gradually, and the peaking part of the

TDS results. However, the correlation between the distance and the displacement fades off with increasing r_{mn}. Referring to eqn (9.6), a limited number of terms describe the scattering from the structure modulated by the acoustic waves. Following the suggestion by Sabine (1988), a number of calculations were made in the case of energy dispersive diffraction. A few examples are shown in Fig. 9.6. It is clear that the modulations of the background due to the TDS can be reproduced, but the number of terms needed in eqn (9.6) is ambiguous. However, it is expected that reliable procedures can be developed through model calculations, where the intensity from eqn (9.6) is compared with that from more precise TDS calculations.

The reasoning leading to eqn (9.6) is equally applicable to disorder scattering, where the modulations arise from short-range correlations of atomic positions or site occupations. In fact, amorphous scattering has been modelled using eqn (9.6) or by the corresponding Fourier filtering (Richardson, Chapter 6).

9.4   Crystallite size, microstrain, and extinction

The effects of crystallite size and microstrain are treated in this chapter only from the viewpoint of Rietveld refinement. For a discussion of the relation of the model parameters to the distributions of crystallite size and microstrain a reference to recent work by Langford et al. (1988) is made. The treatment of size and microstrain effects in this chapter is basically similar to those in Chapters 7 and 8 but brings in some additional aspects. It can be shown on general grounds that the intensity of a reflection falls to zero as the inverse square of the range in the profile tails (Wilson 1962). This is the basis of the so-called variance method for determining the crystallite size (Langford and Wilson 1963). The intensity in the tails may be written as (Suortti and Jennings 1977)

$$i(s) = \frac{\beta_c(s)}{2\pi s^2} \int^s i(s)\, ds, \tag{9.7}$$

where $s = (\cos\theta/\lambda)\, \Delta(2\theta)$, $\Delta(2\theta)$ is the angular deviation of the measured point from the Bragg angle of the reflection, and $\beta_c(s)$ is the integral breadth of the reflection. $\beta_c(s)$ is the inverse of the area-weighted mean thickness τ_A of the crystallites or coherently diffracting domains measured perpendicularly to the diffraction planes,

$$\beta_c(s) = 1/\tau_A \tag{9.8a}$$

$$\beta_c(\Delta(2\theta)) = \lambda/\tau_A \cos\theta. \tag{9.8b}$$

Al, 2θ = 12.2°, 20 nearest neighbours

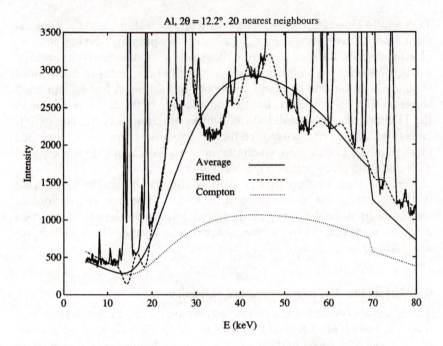

Al, 2θ = 12.2°, 15 nearest neighbours

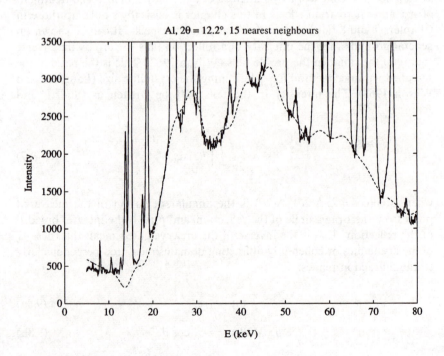

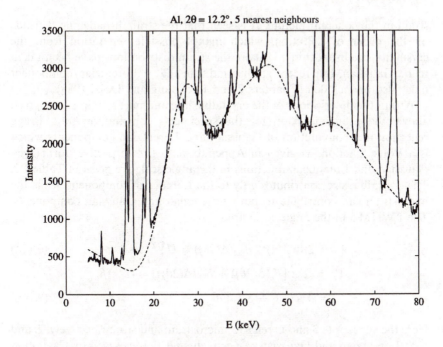

Fig. 9.6 TDS from acoustic phonons in Al calculated from the Debye equation (9.6). The number of the atomic shells of radius r_{mn} is varied in the calculations. The intensity is scaled using the average value of the TDS from eqn (9.3). The contribution of the Compton scattering is also shown.

It is independent of the order of the reflections, although in general it depends on the crystallographic direction. The profile function due to the crystallite size is strictly Lorentzian only far from the peak, but nevertheless, the Lorentzian is used to account for this effect in the whole range of reflection.

The microstrains can be considered to arise either from strain variation within the coherently diffracting domains or from domains with uniform strains which differ within certain limits in the ensemble of incoherently diffracting domains. The interpretation of the strain broadening is quite different in these two pictures, but both give rise to the same angular variation of the integral width of the function describing the line broadening due to microstains β_s namely

$$\beta_s = f(\varepsilon)/2d, \tag{9.9a}$$

$$\beta_s(\Delta(2\theta)) = f(\varepsilon) \tan \theta, \tag{9.9b}$$

where d is the lattice spacing, and $f(\varepsilon)$ is a measure of strain. It is proportional to the average microstrain, $|\Delta d/d|$, in the incoherent picture or to the r.m.s.

strain in the coherent picture. It is seen that the strain broadening depends on the order of reflection, which makes possible separation from the crystallite size broadening. Usually the strain distributions have been taken to be Gaussian, but recent studies indicate that a Lorentzian component must also be included (Chapters 7 and 11; Young and Desai 1989).

Within the descriptions of the crystallite size and microstrain effects given above, the profile function for the specimen's contribution to a Bragg reflection is a convolution of Gaussian and Lorentzian components which is a Voigt function, as given in Appendix 8.A. The respective normalized Gaussian and Lorentzian functions in the angle scale are given in Table 1.2. The crystallite size contributes only to the Lorentzian component, while the microstrain can contribute to both Lorentzian and Gaussian components. The FWHM's in the angle scale are

$$\Gamma_G = 2(\ln 2/\pi)^{1/2}\beta_{SG}(\Delta(2\theta)) = U^{1/2}\tan\theta \tag{9.10a}$$

$$\Gamma_L = (2/\pi)[\beta_c(\Delta(2\theta)) + \beta_{SL}(\Delta(2\theta))] = (2/\pi)\beta_L$$

$$= Y/\cos\theta + X\tan\theta. \tag{9.10b}$$

Here the subscripts S and C refer to microstrain and size respectively, L and G to Lorentzian and Gaussian respectively and U is as given by Caglioti *et al.* (eqn. (1.6)). The integral breadth of the Voigt function resulting from the convolution of these Lorentzian and Gaussian function is

$$\beta_V = \beta_G \exp(-y^2)/[1 - \text{erf}(y)]. \tag{9.11}$$

where $y = \beta_L/\pi^{1/2}\beta_G$ (Langford 1978). Various authors have given approximate forms that relate the integral breadths and FWHM's of the component functions, and thus allow determination of the relative amounts of Lorentzian and Gaussian components (Langford *et al.* 1988).

The above analysis suggests that only three line shape parameters, X, Y, and $U^{1/2}$, are needed in the Rietveld refinement, as long as the distributions of crystallite size and strain can be assumed to be independent of the crystallographic direction. In the case where anisotropy must be included, it is best described with expansions in functions of the correct symmetry properties, such as the spherical harmonic functions that are selected according to the Laue class of the crystal. These functions have been used to model preferred orientation of the sample, and the procedure has been implemented in Rietveld refinement (Ahtee *et al.* 1989). It turns out that only a few functions are needed, and their weights are the parameters determined in the refinement by fitting the integrated intensities. Applied to the case of direction-dependent line shapes, the weights of the functions would be determined by the widths of the reflections. The use of spherical harmonics in modelling preferred orientation has demonstrated that when the effects of anisotropy are large only one or two functions are needed for an adequate description; the most severe effects are easiest to model.

The Lorentzian tails of the reflections extend very far, and there is a loss of intensity due to truncation even when the background is taken at the mid-points between widely separated reflections. The relative loss of the integrated reflection due to broadening is (Suortti and Jennings 1977)

$$(\Delta N_{hkl}/N_{hkl})_b = 2\beta_L(s)/\pi^2 s, \qquad (9.12)$$

where N_{hkl} is the total intensity ($\int i(s)\,ds$) of the reflection. An attempt to minimize this loss by using a sample with large particle size may lead to significant primary extinction effects, however (Chapter 4). If platelet-like particles are assumed, the fractional loss of the integrated reflection can be calculated, and in first order

$$(\Delta N_{hkl}/N_{hkl})_e = [3\Lambda_s \sin \theta \, \beta_L(s)]^{-2}, \qquad (9.13)$$

where $\Lambda_s = V_c/r_0\lambda C F_{hkl}$ is the extinction length including the polarization factor C ($=1$ for the σ polarization, $|\cos 2\theta|$ for the π polarization). It is important that, for a given reflection, this result is approximately independent of the wavelength λ. Calculations indicate that the minimum loss of the integrated reflection in intensity is typically at least 1 per cent and considerably more when all of the crystallites are not of the ideal thickness. It is seen from eqns (9.8) and (9.9) that $\beta_L(s)$ is constant or increases with the magnitude of the scattering vector, $1/d$, while the available scan range is proportional to d. Thus the broadening losses increase with the scattering vector. On the other hand, the extinction depth $\Lambda_s \sin \theta$ increases with $1/d$, and the extinction losses become insignificant. The problem is then reduced to decomposing the pattern into reflections with overlapping Lorentzian tails.

9.5 Conclusions

This work was motivated by the poor performance of the profile shape functions when Rietveld refinement was applied to high-resolution data collected by a conventional X-ray diffractometer (Ahtee *et al.* 1989). Particularly, fitting the model to highly asymmetric reflections was beyond the capacity of the existing codes. In general, it is questionable whether the instrument effects can be modelled adequately by functions involving only a few parameters. Accordingly, a different approach was taken. The instrument function is calculated from the known scattering geometry using ray-tracing. It is natural to include in this calculation all angle-dependent factors of the scattered radiation, the effects of the wavelength and polarization distributions, and the absorption factor. The profile shape function arising from crystallite size and strain is assumed to be Voigtian, where the angular dependencies of the Lorentzian and Gaussian components are described by simple functions involving three parameters. The total profile shape function is found by numerical convolution of the Voigtian by the instrument function, and for each reflection the integral of the profile shape function is weighted by the multiplicity and the square of the structure factor. In addition, a weight function with a few additional parameters may be used to model preferred orientation.

In this approach, the number of profile shape parameters to be found in the course of refinement is reduced to three which are directly related to the distributions of crystallite size and microstrain. This puts the analysis of the profile shapes on a much firmer footing and even allows introduction of anisotropy without making the number of parameters too large. The dependence of the profile width on the crystallographic direction is probably best modelled by spherical harmonics of the symmetry of the crystal, i.e. by the same functions which are used for modelling preferred orientation (Ahtee et al. 1989).

Subtraction of the background is a crucial step in studies of profile shapes. Usually the background is modelled separately, and subtraction procedures may even involve hidden parameters. In the present work, the background is divided into two parts: incoherent and coherent. The first one can be calculated without reference to the model being refined, and the latter one is described with the radial correlation function, which carries the salient features of the TDS and disorder-diffuse scattering. The requirement that the calculated TDS be equal to the part missing from the Bragg reflections due to thermal motion is used as a constraint in separation of the background.

Implementation of the above suggestions will require changes or additions to the Rietveld codes. However, the basic philosophy behind the code will prevail, so the alterations will be of a technical nature only.

9.6 Acknowledgements

The author is grateful for comments and suggestions from Drs M. Ahtee and P. Paatero, and particular thanks are due to Mr V. Honkimäki, who performed most of the calculations. The work was supported financially by the Academy of Finland (projects 01/545 and 10/1042).

References

Ahtee, M., Nurmela, M., Suortti, P., and Järvinen, M. (1989). *J. Appl. Crystallogr.*, **22**, 261–68.

Cox, D. E., Toby, B. H., and Eddy, M. M. (1988). *Austral. J. Phys.*, **41**, 117–31.

Hastings, J. B., Thomlinson, W., and Cox, D. E. (1984). *J. Appl. Crystallogr.*, **17**, 85–95.

Hepp, A. and Baerlocher, Ch. (1988). *Austral. J. Phys.*, **41**, 229–36.

Honkimäki, V. (1990). MSc thesis (in Finnish).

Langford, J. I. (1978). *J. Appl. Crystallogr.*, **11**, 10–14.

Langford, J. I., Delhez, R., de Keijser, Th. H., and Mittemeijer, E. J. (1988). *Austral. J. Phys.*, **41**, 173–87.

Langford, J. I. and Wilson, A. J. C. (1963). In *Crystallography and crystal perfection* (ed. G. N. Ramachandran), pp. 207–22. Academic Press, London.

Sabine, T. M. (1988). Private communication.

Suortti, P. (1967). *Ann. Acad. Sci. Fennicae*, A VI, **240**, 1–33.

Suortti, P. and Freund, A. K. (1989). *Rev. Sci. Instrum.*, **60**, 2579–85.

Suortti, P. and Jennings, L. D. (1977). *Acta Crystallogr.*, **A33**, 1012–27.

Warren, B. E. (1969). *X-ray diffraction*. Addison-Wesley, Reading, MA.

Wilson, A. J. C. (1962). *X-ray optics* (2nd edn). Methuen, London.

Young, R. A. and Desai, P. (1989). *Archiwum Nauki o Materialach*, **10**, 71–90.

10

Restraints and constraints in Rietveld refinement

Christian Baerlocher

10.1   Introduction

In the same year as Rietveld's paper, 'A Profile Refinement Method for Nuclear and Magnetic Structures' appeared (Rietveld 1969), Meier and Villiger (1969) published an article, the German title of which could be translated as 'A Distance Refinement Method for the Determination of Atomic Coordinates of Idealized Framework Structures'. The similarity is evident: both methods refine atomic coordinates by using a least-squares procedure. The only difference is the type of observation. Whereas the former uses diffraction data, the latter uses chemical or geometric observations in the form of known or expected interatomic distances.

The distance least-squares program DLS-76 (Baerlocher et al. 1976) has been widely used for a variety of purposes: to calculate good starting parameters for structure refinement, to screen structural models for unknown phases (e.g. Baur 1977; Meier et al. 1987), and to predict structural changes (e.g. under high pressure) in cases where diffraction data were unavailable (Dempsey and Strens 1976). Despite its simplistic approach, the results obtained are generally remarkably good.

In our earlier work with powder refinement (using integrated intensities), we often wanted to have a means of repositioning atoms which had gone astray in the refinement as a result of the poor observation/variable ratio. Because of the good performance of distance least squares, it was a logical step to include interatomic distances and angles in the Rietveld code as additional observations when the X-ray Rietveld System was written (Baerlocher 1982). A similar approach was used by Hendrickson and Konnert (1980) in another field of crystallography where the over-determination ratio is marginal: the refinement of biological macromolecules.

They incorporated a variety of stereochemical knowledge, ranging from bond distances to torsion angles, to relationships between anisotropic displacement factors, into the refinement process. A further crystallographic application of this technique is the use of least-squares restraints for origin fixing in polar space groups (Flack and Schwarzenbach 1988).

Since one can define about as many such pseudo-observations as there are positional variables, the overdetermination ratio can be significantly increased. The combination has been so successful that it has now been built into a number of Rietveld programs, including GSAS (Larson and von Dreele 1985) and the Cambridge Crystallographic Software Library (CCSL) (David *et al.* 1988). Izumi (Chapter 13) includes an 'exterior penalty function' in his RIETAN program.

10.2 Mathematical methods

In principle, there are many ways in which additional observations or subsidiary relationships as they are also called, can be incorporated into a least-squares refinement process. They can be included as constraints (rigorous or hard constraints) or as restraints (soft or slack constraints). Constraints are imposed rigorously, so the relationship specified in a constraint must be exactly fulfilled. Examples of constraints are the symmetry constraints placed on atoms in special positions, or group constraints, where the distances and angles between atoms within a group are fixed and only the orientation of the group as a whole is refined. Restraints, on the other hand, are relationships which are imposed approximately, and the degree of approximation is given by a finite weight.

Constraints can be imposed in two ways: by the method of Lagrangian multipliers or by the elimination of parameters. The latter is an elegant method since it actually reduces the size of the matrix, but since particular parameters have to be singled out for elimination, a certain asymmetry is introduced. It also requires more cumbersome calculations for both the partial derivatives and the e.s.d.'s and correlation coefficients of the eliminated variables. Of course, an asymmetry is not introduced when new parameters which are a function of the original eliminated variables are defined. An application of this is the group refinement mentioned above. The asymmetry problem can also be avoided by using the method of Lagrangian multipliers, but this method has other disadvantages: the size of the matrix is increased, and special inversion routines are required. Both methods have the general disadvantages of constraints in common: they are inflexible and less versatile than restraints, and they are more cumbersome to compute.

Restraints have many advantages. Above all, they allow deviations from the prescribed values. They are easy to program and compute because they are treated in the same way as a diffraction observation (Waser 1963). Thus, no parameters have to be fixed or eliminated, and the size of the normal equation matrix does not change.

The function minimized in a Rietveld refinement can be written as (Chapter 1)

$$S_y = \sum_i w_i(y_{oi} - y_{ci})^2$$

where y_{oi} and y_{ci} are the observed and calculated intensities at the step i, and $w_i = 1/\sigma^2$, where σ is the standard deviation. Similarly, the function of the restraints or pseudo-observations is written as

$$S_R = \sum w(R_o - R_c(x))^2 \tag{10.2}$$

where R_o can be an expected distance, or angle or any other stereochemical quantity for which an expected value can be obtained, $R_c(x)$ is its value calculated from the atomic positions or other structural variables, and w is again the inverse of the variance of, in this case, the pseudo-observation. For an Si–O distance, which can be assumed to be known to within 0.01 Å, a weight of 10^4 would be taken. Generally, $R_c(x)$ will be a non-linear function, and therefore must be linearized with a Taylor series expansion in the same way as a structure factor equation.

These two minimization functions are then combined as follows:

$$S_{yR} = S_y + c_w \cdot S_R \tag{10.3}$$

where c_w is a common weight factor which can be used to vary the contribution or influence of the restraints in the refinement process. If c_w is made larger, the result will be more and more dominated by the restraints until they act like constraints. On the other hand, c_w can be made smaller as the refinement progresses until the restraints can be removed completely.

Boundary restraints are variants of the restraints (Hepp 1981). Here, a parameter is allowed to move freely within a specified range. As soon as the value of that variable exceeds the boundaries, a penalty function is applied to force it back. Although this approach has some merits in the sense that there is more freedom for the parameters and chances of detecting anomalies in a structure are greater, experience with this type of restraint has shown no advantage over the normal restraints. On the contrary, there was a marked tendency towards oscillation of the parameters.

10.3 How do restraints work?

If one performs a Rietveld analysis combined with restraints, the beneficial effects on the progress of the refinement are obvious to the user. However, it is somewhat difficult to convey the reasons for this to the reader. Since it is a multidimensional problem, it is not easy to illustrate. In order to show how restraints work, a series of plots of the residuals, S_y, S_R, and S_{yR}, as functions of two atomic coordinates are given in Fig. 10.1(a–d).

These plots were calculated using the moderately complex, synthetic, levyne-type zeolite NU-3 (McCusker 1989). The relevant details are given in the table below:

Space group	$R3m$
Cell parameters a	13.118 Å
c	22.554 Å
Radiation	Cu K$_\beta$
FWHM at 35° 2θ	0.18° 2θ
No. reflections	455
No. distance and angle restraints	23
No. atoms	10
No. structural parameters	37

The data are of average quality (as can be judged from the FWHM) for a zeolite. The residuals were calculated at the beginning of the refinement process, when the weighted pattern R-value was still 44 per cent. The coordinates x and z of atom Si(1) were shifted in steps of 0.02 Å in an array of ± 0.3 Å from the final position. All other variables were held constant for these calculations. In order to be able to compare the different residual functions, their maximum values were normalized to 100. The plots shown are typical. Similar plots were also obtained with other sets of parameters.

Figure 10.1(a) shows the rather shallow minimum of the Rietveld residual function. From the contour map at the bottom of the figure, it is evident that the minimum lies more than 0.1 Å away from the correct position. This deviation is also reflected by the fact that a refinement without restraints at this stage gives Si–O distances which vary between 1.40 Å and 1.93 Å. The corresponding residual function of the restraints shown in Fig. 10.1(b) is much steeper, and, moreover, it has its minimum closer to the final position of the atom. The absolute values of the two functions S_y and S_R for this case, range between the following values:

	S_y	S_R	$S_y + 16 \cdot S_R$
minimum	$0.32 \cdot 10^6$	$0.35 \cdot 10^2$	$0.32 \cdot 10^6$
maximum	$0.77 \cdot 10^6$	$0.60 \cdot 10^4$	$0.84 \cdot 10^6$

Compared with S_y, the residual function of the restraints S_R is rather small, but even with that, the refinement is greatly stabilized, and the Si–O distances only vary between 1.59 Å and 1.71 Å. In Fig. 10.1(c), the combined normalized

RESTRAINTS AND CONSTRAINTS

(a)

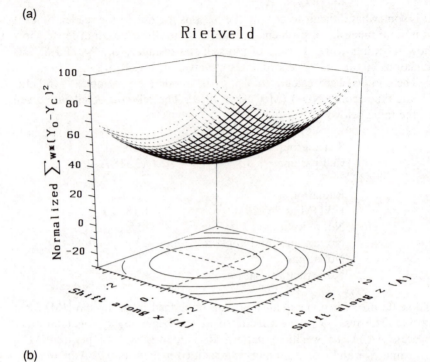

(b)

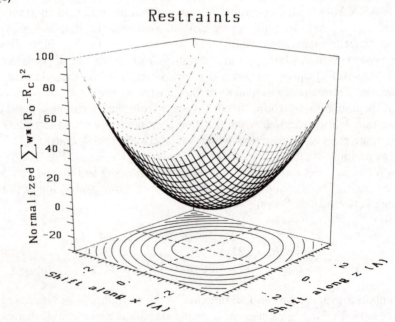

Fig. 10.1 Normalized residuals as a function of two coordinates (see text).

(c)

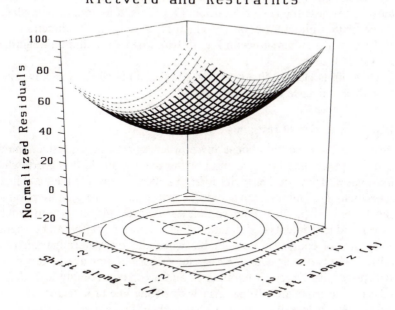

(d)

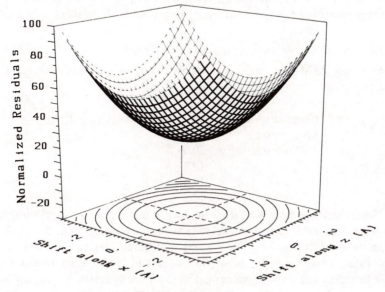

Fig. 10.1   (*continued*)

residual function is plotted with a common weight factor $c_w = 16$. These are only subtle changes, but they are large enough to deepen the minimum and move it towards the correct position. Of course, if a large c_w is applied, the shape of the minimum could be changed entirely in the direction of Fig. 10.1(b). Such a plot is shown in Fig. 10.1(d), where the common weight factor c_w was 144.

These plots clearly illustrate how the restraints work, and how one can tune their influence on the refinement.

10.4   Application of restraints

Restraints are useful in almost any powder refinement, and, since the work and computer time involved in applying them is small, it is worthwhile in most cases. They are beneficial for a number of reasons: they stabilize the refinement, false minima are avoided, convergence is faster, more parameters can be refined (i.e. for more complex structures and in pseudo-symmetry cases), and refinements of diffraction data of inferior quality can be upgraded. They are, of course, most valuable when the overdetermination ratio is critical. The two most common cases for such a situation are (a) very complex structures with large cell parameters and a correspondingly high degree of reflection overlap, and (b) patterns with strong line broadening which gives rise to a severe loss of resolution. To demonstrate the power of the restrained Rietveld refinement, examples of two such refinements are discussed below. Restraints have also been used in conjunction with partial structure solutions (David *et al.* 1989). The known molecule was put into a trial position and strongly restrained so that it moved like a rigid body to its correct location.

10.4.1   *Complex structure of the as-synthesized form of the zeolite ZSM-5*

The zeolite ZSM-5 has a fairly complex structure, which is indicated by the following data

Chemical formula	$[(C_3H_7)_4N]_4Si_{96}O_{192}(OH)_4$
Space group	*Pnma*
Cell parameters a	20.09 Å
b	19.95 Å
c	13.43 Å.

A Rietveld refinement was done to locate the organic cation tetrapropyl-ammonium, TPA (Baerlocher 1984) with conventional Cu $K_{\alpha 1}$ radiation (quartz monochromator). The number of structural parameters was 181, and the total contribution of the organic cation to the number of electrons in the structure was only 10 per cent. The inclusion of restraints was absolutely necessary, and was instrumental in the elucidation of the finer details of the arrangement of the TPA. A section of the difference Fourier map generated with the SiO_2 framework atoms only is presented in Fig. 10.2(a). It shows the weak electron density for two of the four propyl groups and indicates two possible orientations for them. A similar map was obtained by Price *et al.* (1982) using data from a twinned crystal. They could not use any restraints, and their refined C positions formed more or less a straight line (near the average of the two orientations shown in Fig. 10.2(a)) with correspondingly bad distances and angles. With restraints (distances and angles) and powder data, a chemically more acceptable interpretation could be obtained. This can be seen in the final F_0 map in Fig. 10.2(b). Unexpectedly, the propyl groups were found to be oriented in a way different from those in the crystal structure of TPABr (dashed line in Fig. 10.2(a)). This fact was later confirmed by a more detailed refinement with good single crystal data (van Koningsveld *et al.* 1987), where some disorder of the TPA was also detected.

10.4.2 *Transformation of the zeolite ZK-14*

The sodium form of the synthetic zeolite ZK-14 transforms topotactically from a chabasite-type to a sodalite-type phase upon heating (Cartlidge and Meier 1984). This transformation is preceded by large lattice parameter shifts, and a symmetry reduction from $R3m$ to $C2/m$. This causes considerable stress and loss of crystallinity, which produces substantial broadening of the diffraction peaks. Furthermore, the usable 2θ range is reduced, because the peaks at higher 2θ values disappear into the background. Despite the fact that the zeolitic water is lost, and therefore the number of atoms decreases, the reduced symmetry increases the number of parameters from 34 to 50. This loss of X-ray observations coupled with an increase in the number of variables could be somewhat compensated for by the increase (from 14 to 38) in the number of independent distances and angles for the framework atoms resulting from the lower symmetry. This allowed the structure at different stages of the transformation to be refined, and a mechanism to be deduced.

10.5 Discussion

Additional observations are certainly very valuable in a Rietveld refinement, whether they are included as constraints or as restraints. Because of the simplicity, elegance, and flexibility of restraints, this method should be the preferred one. It is well justified to treat the additional information in the same way as the diffraction data. One could, in fact, equally well think of the restrained Rietveld refinement as a distance and angle refinement subject to the additional relationships obtained from a powder pattern. In this

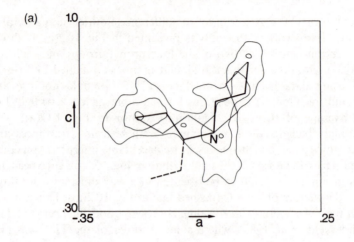

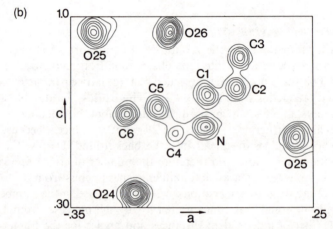

Fig. 10.2 (a) Difference Fourier map of the mirror plane in TPA-ZSM-5 with two possible orientations of the propyl group. (b) Final F_0 map of the same section.

respect, the term 'restrained' refinement is misleading since it declares one type of observation to be superior.

Group refinements (e.g. rigid body or planar molecule) are often performed with constraints. However, it can be argued that this can be done equally well with restraints (D. Schwarzenbach, personal communication). Constraining the appropriate parameters (for example, to maintain the planarity of a molecule) is especially valuable in cases where the diffraction data do not contain sufficient information with respect to those parameters. Under such circumstances, normally weighted restraints are also powerful enough to enforce the required condition. Of course, if the diffraction data contain the relevant information, constraints or restraints are no longer necessary.

It goes almost without saying, that care must be taken when applying restraints. As is true with all data, they have to be measured (selected) carefully. Just as the choice of a wrong space group sometimes remains undetected, false assumptions can also. However, because the influence of restraints can be easily adjusted with the common weight factor, such errors can be minimized.

Still, the question remains: can false assumptions included in the restraint equations be detected? Erroneous assumptions can be looked upon as unknown (systematic or non-systematic) errors in the pseudo-observations. They will have an effect similar to that of such errors in normal data. However, in the case of a restrained refinement, at least one has two completely different types of data, and hence different types of errors. If two data sets contradict one another because of such errors, the refinement will not converge. Thus, data errors are detected more easily in a restrained refinement and non-convergence should therefore serve as a warning. This applies equally well to wrong assumptions in the restraints as to errors in the diffraction data. There is a built-in check on the data used.

10.6 Acknowledgements

I thank Dr A. Hepp for his mathematical contributions during the development of the application of restraints for Rietveld refinement, Prof. D. Schwarzenbach for the informative discussions on restrained refinement and Dr L. B. McCusker for critically reading the manuscript. The work was supported in part by the Swiss National Science Foundation.

References

Baerlocher, Ch. (1982). *XRS-82. The X-ray Rietveld system*. Institut fuer Kristallographie, ETH, Zurich.

Baerlocher, Ch. (1984). *Proceedings of the 6th International Zeolite Conference*, Reno, USA, pp. 823–33.

Baerlocher, Ch., Hepp, A., and Meier, W. M. (1976). *DLS-76, 'a program for the simulation of crystal structures by geometric refinement'*. Institut fuer Kristallographie, ETH, Zurich.

Baur, W. H. (1977). *Phys. Chem. Miner.*, **2**, 3–20.

Cartlidge, S. and Meier, W. M. (1984). *Zeolites*, **4**, 218–25.

David, W. I. F., Akporiaye, D. E., Ibberson, R. M., and Wilson, C. C. (1988). Rutherford Appleton Laboratory Report No. RAL-88-103.

David, W. I. F., Ibberson, R. M., Matsuo, T., Suga, H., Yamamoto, O., and Wilson, C. C. (1989). *International Workshop on the Rietveld Method*, Petten, Abstract A12, p. 43.

Dempsey, M. J. and Strens, R. G. J. (1976). In *Physics and chemistry of minerals and rocks* (ed. R. G. J. Strens), pp. 443–6. Wiley.

Flack, H. D. and Schwarzenbach, D. (1988). *Acta Crystallogr.*, **A44**, 499–506.

Hendrickson, W. A. and Konnert, J. H. (1980). In *Computing in crystallography* (ed. R. Diamond, S. Ramaseshan, and D. Venkatesan), 13.01–26. Indian Academy of Sciences, Bangalore.

Hepp, A. (1981). Ph.D. Thesis, University of Zurich.

Larson, A. C. and von Dreele, R. B. (1985). Los Alamos National Laboratory Report No. LA-UR-86-748.

McCusker, L. B. (1989). Recent Research Reports, *8th International Zeolite Conference*, Amsterdam, Holland, pp. 281–2.

McCusker, L. B., Meier, W. M., and Rechsteiner, H. (1987). *Mater. Res. Bull.*, **22**, 1203–7.

Meier, W. M. and Villiger, H. (1969). *Z. Kristallogr.*, **129**, 411–23.

Price, D. G., Pluth, J. J., Smith, J. V., Bennett, J. M., and Patton, R. L. (1982). *J. Am. Chem. Soc.*, **104**, 5971–7.

Rietveld, H. M. (1969). *J. Appl. Crystallogr.*, **2**, 65–71.

van Koningsveld, H., van Bekkum, H., and Jansen, J. C. (1987). *Acta Crystallogr.*, **B43**, 127–32.

Waser, J. (1963). *Acta Crystallogr.*, **16**, 1091–4.

11

Rietveld refinement with time-of-flight powder diffraction data from pulsed neutron sources

W. I. F. David and James D. Jorgensen

11.1 Introduction

The recent development of accelerator-based pulsed neutron sources has led to the widespread use of the time-of-flight technique for neutron powder diffraction. The properties of the pulsed source make possible unusually high resolution over a wide range of d-spacings, high count rates, and the ability to collect complete data at fixed scattering angles. The peak shape and other instrument characteristics can be accurately modelled, which makes Rietveld refinement possible for complex structures. In this chapter we briefly review the development of the Rietveld method for time-of-flight diffraction data from pulsed neutron sources and discuss the latest developments in high resolution instrumentation and advanced Rietveld analysis methods.

11.2 Early history of the time-of-flight Rietveld method

The first Rietveld refinement using time-of-flight neutron powder diffraction data was performed in 1974 to analyse the monoclinic phase of KCN from data taken in a high pressure cell (Decker *et al.* 1974). This experiment was part of an extended program using the time-of-flight technique with a neutron chopper on a reactor source, to study the structures of materials at high pressure. The time-of-flight technique was chosen for this work because it made possible the collection of data at a fixed scattering angle for which the scattering from the pressure cell could be excluded by viewing the scattered neutrons only through a narrow window in the cell.

The time-of-flight technique proved ideal for structural studies at high pressure and a number of experiments (Worlton and Beyerlein 1975; Jorgensen et al. 1978a; Jorgensen 1978; Cartz et al. 1979, Jorgensen and Clark 1980; Cartz and Jorgensen 1981; Jorgensen et al. 1984) were performed between 1974 and 1979 using two different time-of-flight diffractometers at Argonne's CP-5 reactor. In all of these cases the data were analysed by the Rietveld method.

The first analysis code for time-of-flight data was written at Argonne National Laboratory and incorporated features that specifically addressed the problems encountered in high pressure diffraction experiments (Worlton et al. 1976). For example, the code was capable of fitting up to four phases. This feature was included so that data from a pressure calibrant such as CsCl could be analysed along with the sample data. The sample pressure was then accurately determined from the refined lattice constant for the calibrant. These first codes for Rietveld analysis of time-of-flight data were not actually based on the constant-wavelength code developed by Rietveld (1969) (and were, in fact, written with no knowledge of Rietveld's earlier work). There were three important differences. First, and most obvious, the extrinsic variable was time-of-flight, not scattering angle. Second, the variation of the resolution function with time-of-flight could be described by a simple function,

$$R(d) = \Delta d/d = (A + Bd^2)^{1/2}, \tag{11.1}$$

where d is the d-spacing, which is linearly proportional to the time-of-flight, t, and A and B are instrumental constants. Third, since the flux on the sample is a function of wavelength, the calculated intensity model included a wavelength-dependent term for the incident flux.

11.3   Time-of-flight diffractometers at reactor sources

Although the resolution of the early reactor-based time-of-flight diffractometers was not high by present standards ($\Delta d/d \geq 0.006$) (Worlton et al. 1976), the peak shape function was almost perfectly Gaussian, as is shown in Fig. 11.1. This ability to model accurately the peak shape and its wavelength dependence undoubtedly led to the success of the Rietveld method for this application. However, except for high pressure structural studies, the time-of-flight technique did not enjoy widespread use. One of the reasons is that, for a chopper-based diffractometer, the variation of the instrumental resolution with wavelength is opposite to that which is most useful, i.e. the resolution is worst at short d-spacings where the heaviest peak

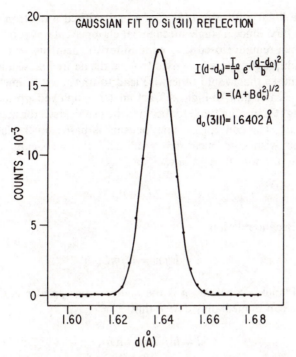

Fig. 11.1 Comparison of a measured diffraction peak for the chopper-based time-of-flight diffractometer at Argonne's CP-5 research reactor with a calculated Gaussian.

overlap occurs. This instrument property arises directly from the way the neutron pulses are formed. With a chopper, the width of the pulse, Δt, in time-of-flight, t, is constant. Thus the resolution, $\Delta d/d = \Delta t/t$, is simply proportional to $1/d$.

In spite of this limitation, the time-of-flight technique can be used to achieve unusually high resolution by collecting the complete diffraction pattern in back scattering. The overall instrumental resolution function has the general form (Worlton *et al.* 1976):

$$R(d) = \Delta d/d = [(\Delta t/t)^2 + (\Delta \theta \cot \theta)^2 + (\Delta L/L)^2]^{1/2} \qquad (11.2)$$

where θ is the scattering angle and L is the total path length from the point at which the pulses are formed to the sample and then to the detector. High resolution can be achieved in a straightforward way by placing the detector at 2θ approaching $180°$ and making the path length, L, long, which also lengthens the overall time-of-flight, t.

Steichele and Arnold (1973) demonstrated this concept with a high resolution, back-scattering time-of-flight diffractometer at the Garching reactor. The flight path was 145 m long and utilized a neutron guide tube to maintain flux. Since a single mechanical chopper cannot produce a short pulse and then remain closed long enough for the neutrons to travel such a distance, multiple choppers were used to form the desired pulse and eliminate additional pulses that would otherwise lead to unwanted 'frame overlap'.

All of these early time-of-flight diffractometers employed a principle called time focusing (Carpenter 1967) to increase the usable detector area, and thus the count rate. The concept of time focusing is based on the fact that the variable one wishes to measure is actually the d-spacing, d, not the time-of-flight, t. From Bragg's law,

$$\lambda = 2d \sin \theta, \tag{11.3}$$

and the de Broglie relation,

$$\lambda = h/mv = ht/mL, \tag{11.4}$$

where h is Planck's constant, m is the neutron mass, and v is the neutron velocity, one can immediately derive that

$$d = ht/(2mL \sin \theta), \tag{11.5}$$

and it becomes apparent that if the detectors are placed on a locus defined by

$$L \sin \theta = \text{constant}, \tag{11.6}$$

the neutrons scattered at different angles from the same d spacing will be detected at the same time-of-flight. Since L in eqn (11.6) is the total path length, the desired detector arrangements are not physically realizable in all cases (Jorgensen and Rotella 1982; Jorgensen et al. 1989a). For example, if the incident flight path (source to sample) is much longer than the scattered flight path (sample to detector), as was the case for the high resolution instrument of Steichele and Arnold (1973), the multiple detectors can be placed on the time-focused locus only in back scattering (where the sine function is slowly varying).

11.4 Time-of-flight diffractometers at pulsed sources

Many of the shortcomings associated with time-of-flight diffraction on reactor sources have been overcome by advanced instrument designs implemented on the pulsed neutron sources (Windsor 1981; Brown et al. 1982; Jorgensen and Rotella 1982; Jorgensen et al. 1989a). The ability to form neutron pulses without a chopper provides several advantages. For example,

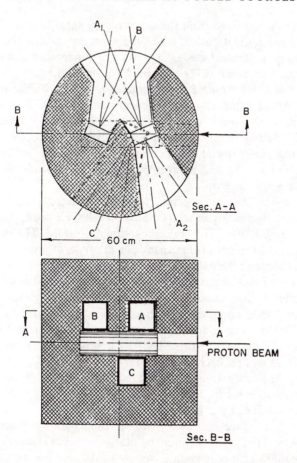

Fig. 11.2 Arrangement of the spallation target and moderators of the Intense Pulsed Source Neutron Source at Argonne National Laboratory (from Carpenter 1977; Carpenter *et al.* 1978). High energy protons from the accelerator strike a cylindrical uranium target. The high energy neutrons thus produced in the target by spallation are moderated to lower energies (typically room temperature or below) in three hydrogenous moderators (A, B, and C) surrounding the target. Because of the relatively small dimensions of the moderators, a fraction of the neutrons in the final beam are undermoderated (i.e. possess energies higher than the moderator temperature). Neutron instruments view the surfaces of these moderators.

the effective source area can be much larger than can be obtained straight-forwardly with choppers. Additionally, the ratio of the pulse width to the time between pulses can be optimized.

At a pulsed source, the pulses are produced when a short burst of high energy neutrons produced by spallation in the target are moderated to thermal energies in a nearby moderator of comparably small dimensions, as shown in Fig. 11.2 (Carpenter 1977; Carpenter *et al.* 1978). The initial high-energy neutron burst from the target is very short (less than 1 μs). The thermal neutron pulse duration, which is longer, depends on the moderator temperature and physical design and on the moderated neutron energy (Graham and Carpenter 1972). One might estimate the pulse width as the distribution of times required for thermal neutrons to diffuse to the front surface of the moderator. Thus, higher energy neutrons exhibit shorter pulses. To first order, the pulse width, Δt, is proportional to wavelength, or equivalently, $\Delta t/t$ is nominally constant. Because of this, the undesirable resolution characteristics of chopper-based spectrometers are largely overcome. Furthermore, the time between pulses can be independently controlled to eliminate 'frame overlap'.

The limitations in achieving time focusing by detector placement have been overcome by a newly-developed method called electronic time focusing (Jorgensen et al. 1989a). The concept is straightforward. The detectors are placed in any convenient arrangement (for example, at a constant distance from the sample) and the data for each detector are separately time encoded. These individual detector data are then combined into a single histogram by computing the flight times for each detector that correspond to the same d-spacing and then combining the appropriate (constant d-spacing) time channels. This focusing computation can either be done in real time (while the data are being collected) or later as a separate step. The real-time electronic focusing approach was used for the time-of-flight powder diffractometers at Argonne's Intense Pulsed Neutron Source (Jorgensen et al. 1989a). These instruments cover a wide range of d-spacings by employing detectors at various angles between 12° and 157°.

A similar procedure has been used at the High Resolution Powder Diffractometer, HRPD, at ISIS for the past three years (David et al. 1988). At ISIS the diffraction patterns are collected with time bin widths that vary in a logarithmic manner as a function of time-of-flight (i.e. $\Delta t = \alpha t$). Typically, on HRPD and POLARIS (a medium-resolution, high-intensity powder diffractometer at ISIS) Δt(bin width) $= 10^{-4}t$ and $10^{-3}t$ respectively. This logarithmic time-binning scheme has two significant advantages: first, over the whole diffraction pattern each Bragg peak is spanned by roughly the same number of time bins, and, second, time focusing can be achieved simply by including a channel-number offset for each detector in the computer memory, i.e. a simple register shift permits real-time detector focusing.

11.5 Rietveld refinement with pulsed neutron source data

Although the development of pulsed neutron sources motivated and, in some ways, made possible the development of advanced time-of-flight diffraction techniques, the characteristics of the pulsed source presented some challenging problems in applying the Rietveld method. The most important is the peak shape and its wavelength dependence. The neutron pulse from the moderator is highly asymmetric in time. The leading edge is very sharp, because the first neutrons to emerge are almost coincident with the high energy neutron pulse hitting the spallation target, while the trailing edge decays according to the moderator size and temperature. For most instrument designs, the other contributions to the resolution function are nominally Gaussian. Thus, the overall peak shape is the convolution of a Gaussian term with a function that describes the initial neutron pulse, and is highly asymmetric and non-Gaussian.

Windsor and Sinclair (1976) used a two-parameter asymmetric Gaussian peak profile to obtain reasonably good fits for nickel powder data from a pulsed source at the Harwell Linac. Later, analysis of high resolution nickel data on a 14 m backscattering diffractometer required a more complex peak shape function based on a Gaussian leading edge and a second Gaussian trailing edge with an exponential tail (Windsor et al. 1980; Cole and Windsor 1980). An entirely empirical approach was demonstrated by Mueller et al. who used a tabulated numerical peak shape function to fit data for Th_4D_{15} from the ZING-P pulsed neutron source at Argonne (Mueller et al. 1977; Jorgensen et al. 1978b).

These first attempts at applying the Rietveld method to spallation pulsed neutron source data demonstrated the feasibility of the technique, but were difficult to apply and obscured the basic physics leading to the unusual peak shape. The first peak shape function that enjoyed widespread use in time-of-flight Rietveld codes was that proposed by Jorgensen et al. (Jorgensen et al. 1978b; Carpenter et al. 1975), based on a convolution of separate rising and falling exponentials that represented the time dependence of the initial neutron pulse with a symmetric Gaussian term that represented the other contributions to the peak shape. The integrals can be done in closed form and the resulting peak shape function has the form:

$$I(t) = \alpha\beta[\exp(u)\,\mathrm{erfc}(y) + \exp(v)\,\mathrm{erfc}(z)]/[2(\alpha + \beta)], \qquad (11.7)$$

where $\mathrm{erfc}(y) = 1 - \mathrm{erf}(y)$ and $\mathrm{erf}(y)$ is the error function, and

$$u = \alpha(\alpha\sigma^2 + 2t)/2, \qquad v = \beta(\beta\sigma^2 - 2t)/2,$$
$$y = (\alpha\sigma^2 + t)/(2^{1/2}\sigma), \qquad z = (\beta\sigma^2 - t)/(2^{1/2}\sigma). \qquad (11.8)$$

Terms α and β characterize the rising and falling exponentials used to model the time dependence of the neutron pulse (with respect to a reference time

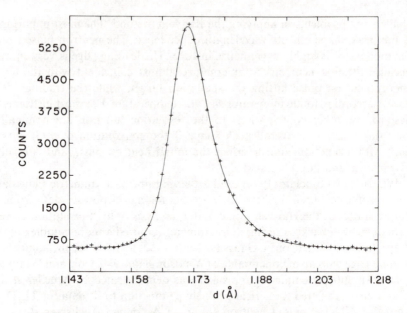

Fig. 11.3 Comparison of the 211 Bragg peak of iron, measured at Argonne's ZING-P spallation neutron source, with the calculated function of eqn (11.7). (From Carpenter *et al.* 1975; Jorgensen *et al.* 1978*b*.)

t^* that corresponds to the maximum of the pulse),

$$\exp(\alpha t^*), \quad t^* < 0 \quad \text{and} \quad \exp(\beta t^*), \quad t^* > 0, \qquad (11.9)$$

and σ describes the width of the Gaussian component.

With α, β, and σ as adjustable parameters, this function provided a precise fit to the observed peak shapes, as shown in Fig. 11.3 (Carpenter *et al.* 1975; Jorgensen *et al.* 1978*b*; Albinati and Willis 1982). The incorporation of this peak shape function into a Rietveld code, however, required that the wavelength (or d-spacing) dependence of α, β, and σ be specified. This information was obtained from diffraction data for standard samples and the required dependence empirically fitted to analytical functions by Von Dreele *et al.* (1982) to produce the first widely used pulsed-source time-of-flight Rietveld code.

Subsequent codes are mostly based on this same peak shape function, sometimes with minor modifications, and have been very successful. However, a recent code written by Izumi *et al.* (1982) uses a more complex peak shape function, the Ikeda–Carpenter function (Ikeda and Carpenter 1985), that provides an improved fit to the peak shapes at small d-spacings or where cold moderators are used because it models more accurately the time

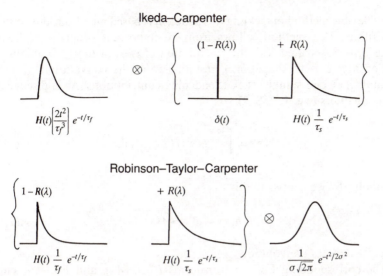

Fig. 11.4 Schematic illustration of the functions that are convoluted to produce the Ikeda–Carpenter and Robinson–Taylor–Carpenter time-of-flight peak-shape functions.

dependence of the initial neutron pulse under such conditions. This peak shape and the related Robinson–Taylor–Carpenter peak shape described in the next section have proved to be successful in high resolution studies at ISIS.

11.6 The peak shape in high resolution time-of-flight powder diffraction data

The Ikeda–Carpenter (IC) function (Ikeda and Carpenter 1985) represents the state-of-the-art description of the neutron pulse shape emanating from a moderator at a pulsed neutron source. It is a rather complicated function consisting of epithermal (slowing down) and thermal (storage term) components. The function is illustrated in Fig. 11.4 along with another popular peak shape description ascribed to Robinson, Taylor, and Carpenter (RTC). Although the Ikeda–Carpenter function more appropriately describes the short-wavelength epithermal region, both functions oversimplify the reflection-profile shape function in the 'switching' region. However, for the purposes of profile fitting in Rietveld refinement, these line shapes are adequately precise.

The computation of the IC and RTC functions is easily tractable with modern day computers. However, with the advent of high resolution time-of-flight powder diffractometers, the folding of sample broadening contributions into the instrumental line shape must be accounted for. The most flexible method of tackling this problem is based on a Fourier transform algorithm. This method follows from an important property of Fourier transforms associated with the convolution of two (or more) functions.

Consider a peak shape, $h(x)$, resulting from the convolution of instrumental, $g(x)$, and sample $f(x)$, broadening contributions. As is discussed in other chapters (e.g. 1, 7, 8, 9),

$$h(x) = \int_{-\infty}^{\infty} g(x')f(x - x') \, dx'. \tag{11.10}$$

Symbolically we can write this as

$$h(x) = g(x) * f(x). \tag{11.11}$$

where $*$ denotes convolution.

The corresponding Fourier transforms $H(k)$, $G(k)$, and $F(k)$ are simply related by the product equation

$$H(k) = G(k)F(k). \tag{11.12}$$

For complete convolutions, the product rule for Fourier transforms also holds; i.e. if

$$h(x) = g_1(x) * g_2(x) * \cdots * g_n(x) * f_1(x) * f_2(x) * \cdots * f_m(x), \tag{11.13}$$

then

$$H(k) = G_1(k)G_2(k)\cdots G_n(k)F_1(k)F_2(k)\cdots F_m(k). \tag{11.14}$$

(See Chapters 7 and 8 for applications of these Fourier transform relations to profile shapes in constant wavelength X-ray data.) A good example of the success of the Fourier transform approach to peak shape description in TOF neutron data is provided by the study of oxygen loss in $YBa_2Cu_3O_{7-x}$ (David et al. 1989). $YBa_2Cu_3O_{7-x}$ possesses a wide range of stoichiometry for $x \approx 0$ (92 K superconductor) to $x \approx 1$ (insulating antiferromagnet). Samples of nominal composition $x = 1$ were prepared by heating cold-pressed slabs of $YBa_2Cu_3O_{6.9}$ at 850°C *in vacuo* (10^{-4} Torr) for 12 h. Diffraction data recorded on HRPD at ISIS indicated that the material had transformed to the anticipated tetragonal structure. Standard profile refinement was performed using a Voigt $*$ double-exponential peak shape function. However, the attempts at refinement were only modestly successful because of a pronounced anisotropic, asymmetrical broadening associated with the

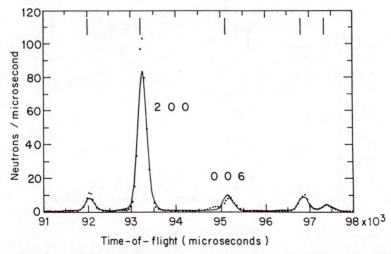

Fig. 11.5   Observed and calculated profiles for a portion of the time-of-flight powder diffraction data for $YBa_2Cu_3O_{7-x}$. The crystallographically isotropic line-shape model gives a poor fit, particularly for 001 reflections.

c axis (Fig. 11.5). Given that the c axis expands substantially on removal of oxygen, it was hypothesized that this broadening was caused by an oxygen gradient within the slab with the lowest oxygen content on the surface of the slab. Assuming linear variations of c vs oxygen stoichiometry and oxygen content vs depth into the slab, the effective scattering density for the resulting c-axis microstrain distribution may be expressed as a truncated quadratic equation, $y = 3t_\Delta^2/\eta^3$ for $0 < t_\Delta < \eta$ and $y = 0$ elsewhere, where η is the time range over which the scattering extends and t_Δ is the time-of-flight within this range. Note that this η which is a function of hkl and is maximum along c and zero for $hk0$ reflections, is not to be confused with the use of η in Chapter 1 and others, meaning the mixing parameter in the pseudo-Voigt profile functions. The precise formula is given by

$$\eta = t_0(\Delta c/c)(ld/c)^2, \tag{11.15}$$

where t_0 and d are the time-of-flight and d-spacing associated with the hkl reflection, c and $\Delta c/c$ are, respectively, the c-axis and fractional shift in c-axis, and l is the third Miller index. Accordingly this additional functional contribution was folded into the existing Voigt $*$ double exponential function and Rietveld refinement performed. The improved fit is evident from a comparison of Figs 11.5 and 11.6. Chi-squared improves from 4.59 to 1.69: the weighted profile R-factor decreases from 14.2 per cent to 8.5 per cent against an expected R-factor of 6.6 per cent. Most importantly, the refined

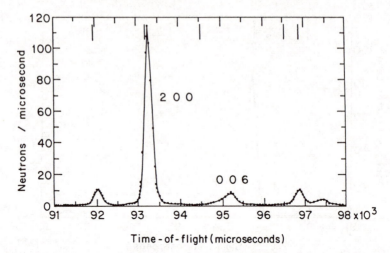

Fig. 11.6 Observed and calculated profiles for a portion of the time-of-flight powder diffraction of $YBa_2Cu_3O_{7-x}$. The line-shape model containing the convoluted anisotropic truncated quadratic function (see text) gives an excellent profile fit for all reflections.

average oxygen content changes from 6.038(8) to 6.075(5) with an estimated variation, from the width of the truncated quadratic function, of ~ 0.2 formula units of oxygen across the slab. This result not only highlights the complexity of peak shape that is necessary for successful Rietveld refinement of high resolution data, but also the care required in preparation of oxygen-deficient samples of $YBa_2Cu_3O_{7-x}$.

11.7   Anisotropic line-broadening in time-of-flight powder diffraction data

Although of no particular benefit in many Rietveld refinement studies, the roughly constant $\Delta d/d$ resolution of a time-of-flight powder diffractometer is particularly useful in the study of structural phase transitions and the analysis and separation of crystallite size and microstrain broadening.[†] In the former case, the advantage of constant $\Delta d/d$ resolution rests in the fact that all orders of reflection are split by the same amount. With the large range of available d-spacings it is not uncommon that splittings are observed in 3 or 4 orders of a given reflection. This permits a uniquely detailed description of structural distortions and pseudosymmetry. For example, two different measurements of the monoclinic angle of NiO at 90.05953(4)° and 90.06051(6)° on HRPD (ISIS) were accounted for by the differences in ambient temperature during the experiment (15 °C and 23 °C). In the case of size/strain effects, the

---

[†] See Chapters 7, 8, and 9 for discussion of crystallite size and microstrain broadening in the context of Rietveld refinement with constant wavelength X-ray data.

excellent resolution at long d-spacings permits the easy discrimination of size and strain effects. For crystallite size effects, it is easily shown that

$$\Delta d^* = \Delta d/d^2 \sim 1/\tau, \qquad (11.16)$$

where P = crystallite size, i.e. $\Delta d/d \sim d/\tau$.

By contrast, for microstrain, $\langle \varepsilon \rangle$,

$$\langle \varepsilon \rangle = \Delta d/d = \Delta d^*/d^*. \qquad (11.17)$$

Thus, peak widths vary linearly with time-of-flight for sample microstrain and the instrumental resolution function, and quadratically for crystallite size effects. These contrasting width variations can be extremely pronounced when several orders of reflection are available.

With the advent of high resolution instrumentation such as the HRPD at ISIS, sample broadening in general dominates the intrinsic instrumental line shape. A version, SAPS (Structure And Peak Shape refinement), of the Rietveld technique has been developed at ISIS to analyse peak broadening in a model-independent manner (David *et al.* 1988). Instead of accounting for the peak width by using a smooth functional variation that is parameterized in time-of-flight, the Gaussian and Lorentzian components of each peak may be separately refined. This results in a multi-parameter Rietveld problem. To ensure convergence, the widths of weak peaks may be constrained or the widths of closely overlapping peaks may be refined together. This approach often leads to substantial improvements in the Rietveld refinement of high resolution data and has the distinct advantage that no assumptions are made about a particular line-broadening model. Similar procedures applied with constant wavelength X-ray data are taken up in Chapters 7, 8, and 9.

A good example of the use of this version of the Rietveld procedure is provided by the analysis of data for $LaNbO_4$ collected on the HRPD at ISIS (David 1990). High resolution neutron powder diffraction data were collected at a series of temperatures from a $1\,cm^3$ sample of $LaNbO_4$. Each run lasted for approximately 30 min. A small portion of the range of d-spacings surveyed is shown in Fig. 11.7. There is clear evidence of a structural phase transition. Indeed, the refined lattice parameters indicate a large monoclinic distortion that collapses in a mean-field manner at the transition temperature. The monoclinic shear angle is displayed in Fig. 11.8.

Examination of the diffraction data collected at temperatures near the phase transition reveals significant broadening, particularly of some $hk0$ reflections. Standard Rietveld refinement with data obtained below 180 °C and above 540 °C proceeded routinely. However, at intermediate temperatures the refinements became progressively worse as the sample temperature

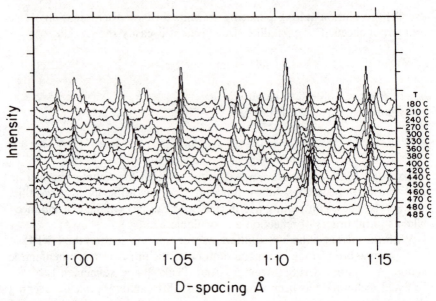

Fig. 11.7 Diffraction patterns from a $1\,cm^3$ sample of $LaNbO_4$ recorded as a function of temperature. The peak splitting associated with the ferroelastic phase transition is evident.

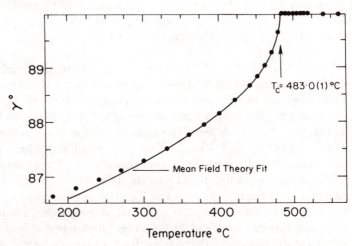

Fig. 11.8 The variation of the monoclinic angle of $LaNbO_4$ as a function of temperature. Least-squares fitting confirms a mean-field behaviour and a transition temperature of 483.0(1) °C.

approached that of the phase transition. Indeed, differences between diffraction patterns recorded at room temperature and the transition temperature were extremely pronounced. Using SAPS, the peak widths were refined as independent variables. Structural parameters for $LaNbO_4$ that were unbiased by an imposed peak broadening model were thus obtained. A selection of the (Lorentzian) peak width components so refined are presented in Table 11.1.

These peak widths were then analysed to determine the origin of the line broadening. Refinements at various temperatures revealed a pronounced temperature dependence that diverges around T_c. The anomalous Lorentzian broadening of the 220, $\bar{2}20$, and 112, $\bar{1}12$ reflections are displayed in Fig. 11.9. The ratio of the slopes above and below T_c is 2:1 within experimental error, as predicted by mean-field theory. The predominant nature of the Lorentzian broadening is consistent with microstrain effects described by a second rank tensor. Accordingly, anisotropic Lorentzian strain and particle size effects described by a second rank tensor formulation were added in convolution to the standard time-of-flight peak shape. The full-width-at-half-

Table 11.1   Part of one of the output files from SAPS from the refinement with data for $LaNbO_4$ near the phase transition

h	k	l	I	$\sigma(I)$	Γ	$\sigma(\Gamma)$	σ	$\sigma(\sigma)$
0	1	5	663.8	20.5	25.3	3.9	51.8	0.0
1	0	5	540.2	20.6	43.1	5.3	51.9	0.0
−1	1	4	3646.5	37.8	49.7	1.7	55.9	0.0
1	1	4	3460.5	38.5	45.5	1.7	56.2	0.0
−1	2	1	715.1	24.2	112.5	8.5	56.7	0.0
−2	1	1	2443.0	31.0	40.7	2.1	57.1	0.0
1	2	1	2192.0	28.3	38.5	2.2	57.3	0.0
2	1	1	836.2	26.5	97.4	6.6	57.7	0.0
0	2	2	461.3	29.5	89.7	10.3	59.0	0.0
2	0	2	579.9	30.7	58.2	6.5	59.6	0.0
0	2	0	10.3	35.4	42.2	0.0	64.9	0.0
−2	0	0	0.7	35.7	42.8	0.0	65.8	0.0
0	0	4	121.4	43.7	45.9	0.0	70.6	0.0
0	1	3	361.6	45.4	49.6	0.0	76.2	0.0
1	0	3	303.1	30.3	49.8	0.0	76.5	0.0
−1	1	2	9352.3	89.6	114.4	2.7	77.0	0.0
1	1	2	10061.0	98.0	93.5	2.3	77.7	0.0

The table contains Miller indices, *hkl*, along with peak intensities, *I*, and standard deviations, $\sigma(I)$. If $I/\sigma(I)$ is greater than a predetermined value (12 in this particular case), peak widths are refined. For $I/\sigma(I) < 12$, peak widths are fixed (note zero standard deviations for some Γ). In this example the usual TOF functional form for σ was refined. All individual Gaussian widths were fixed.

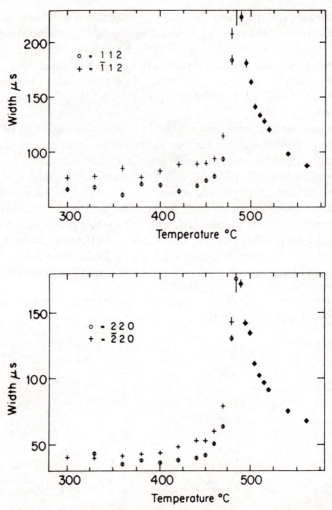

Fig. 11.9 Refinement of individual peak widths for LaNbO$_4$ reveals a pronounced temperature dependence that diverges around the phase transition temperature, T_c. The distinction between 220 and $\bar{2}$20 (similarly, 112 and $\bar{1}$12) peak widths is consistent with strain effects described by a second rank tensor.

maxima formulae used were

$$\Gamma_{\text{strain}} = \xi d^2; \qquad \Gamma_{\text{size}} = \xi d^3 \qquad (11.18)$$

where

$$\xi^2 = \Gamma_{11}h^2a^{*2}\Gamma_{22}k^2b^{*2} + \Gamma_{33}l^2c^{*2} + 2[\Gamma_{23}klb^*c^* \cos \alpha^* + \Gamma_{13}hla^*c^* \cos \beta^*$$
$$+ \Gamma_{12}hka^*b^* \cos \gamma^*]. \qquad (11.19)$$

Table 11.2 Refined anisotropic strain broadening terms for $LaNbO_4$ vs temperature

T (°C)	Γ_{11}	Γ_{22}	Γ_{12}
240	205 (15)	402 (20)	600 (300)
270	198 (15)	366 (20)	900 (300)
300	183 (15)	392 (20)	1600 (300)
330	227 (20)	378 (20)	1500 (300)
360	190 (15)	447 (25)	2500 (400)
380	214 (20)	462 (25)	2300 (400)
400	257 (20)	417 (25)	2600 (500)
420	266 (20)	598 (30)	3900 (600)
450	400 (25)	676 (30)	6500 (1000)
460	460 (40)	770 (40)	8700 (1500)
470	840 (40)	1320 (50)	15 300 (3000)
480	2200 (200)	3800 (250)	36 000 (5000)

Rietveld refinement at a number of different temperatures indicated that the anisotropic broadening results from microstrain rather than size effects. This anomalous behaviour is restricted to the *ab* plane; no extra broadening is associated with the *c*-axis. The refined values shown in Table 11.2 clearly indicate microstrain anisotropy within the *ab* plane.

The anisotropic broadening of the Bragg peaks (microstrain) in $LaNbO_4$ may, as with the monoclinic lattice distortion (macrostrain), be described in terms of a second rank tensor. The principal axes of the ellipsoid representing the microstrain broadening is rotated by 21° from *a**. This agrees to within experimental error ($\sim 2°$) with the orientation of the principal axes of the spontaneous macrostrain hyperbola for all data sets collected between $(T_c - 50°)$ and T_c.

The coincidence of the orientation of microstrain (obtained from line broadening considerations) and macrostrain (calculated from peak splittings associated with monoclinic symmetry) was unexpected and is at present unexplained, although the microstrain broadening is probably associated with crystal imperfections such as dislocations. For the purposes of the present paper it highlights the immense amount of detail that may be derived from the Rietveld refinement with powder diffraction data when sophisticated models for the peak broadening are employed.

11.8 Wavelength dependent effects in time-of-flight Rietveld refinement

The total number of neutrons, i.e. the integrated intensity, of a Bragg reflection measured in a time-of-flight powder diffraction experiment is given by Buras and Gerward (1975):

$$I_{hkl} = \frac{I_0(\lambda)\varepsilon(\lambda)\lambda^4 V_s A_{hkl}(\lambda) E_{hkl}(\lambda) j |F_{hkl}|^2 \cos \Delta\theta}{4V_c^2 \sin^2 \theta}. \tag{11.20}$$

The formula holds for the full Debye–Scherrer cone. The notation used is as follows:

$I_0(\lambda)$ = incident neutron flux at wavelength λ,
$\varepsilon(\lambda)$ = detector efficiency at wavelength λ,
$A_{hkl}(\lambda)$ = attenuation coefficient for reflection hkl,
$E_{hkl}(\lambda)$ = extinction coefficient for reflection hkl,
V_s = sample volume,
V_c = unit cell volume,
j = reflection multiplicity,
F_{hkl} = structure factor,
2θ = Bragg angle.

The first four symbols in the above list are wavelength dependent and must be correctly evaluated for a precise Rietveld refinement to be obtained. The product $I_0(\lambda)\varepsilon(\lambda)$ is usually determined by measuring the incoherent scattering from vanadium and accounting for inelasticity and multiple scattering effects. Although this is a non-trivial calculation that depends on the vanadium sample geometry (which should be identical to the sample configuration), cross-calibration with in-beam monitors is sufficient to allow normalization to proceed using corrected monitor spectra. The analysis of extinction and attenuation effects as a function of wavelength is more complex and requires detailed consideration. However, Sabine (Chapter 4 and Sabine 1988) and Sabine *et al.* (1988) have recently shown that severe extinction can be correctly modelled (if anisotropic) and, therefore, is not an impediment to time-of-flight Rietveld refinement. Similarly, attenuation (including absorption, self-attenuation, and multiple scattering) may be refined to good precision in most cases using a simplified attenuation description given by Sabine (Chapter 4).

The combined refinement of scale factor, temperature factors, attenuation, and extinction may lead to correlated or unstable Rietveld refinements. In refinements where accurate temperature factors are required, the experimental evaluation of sample attenuation using a combination of monitors before and after the sample is essential. For the case of a slab-shaped sample, the sample attenuation may be shown, to a good approximation, to be given by the formula

$$A(\lambda) = [R^2(\lambda) - 1]/[2 \ln(R(\lambda))] \tag{11.21}$$

where $R(\lambda)$ is the ratio of neutron counts in the downstream and upstream monitors.

11.9 Precision and accuracy with time-of-flight Rietveld refinement

The high resolution and large accessible range of d-spacings (particularly short d-spacings) in a time-of-flight powder diffraction measurement, in principle permit the determination of atomic coordinates and temperature factors in moderately complex crystal structures with both high precision, i.e. small standard deviations, and high accuracy, i.e. agreement with the physically correct values (as established by independent measurements). Assuming a constant $\Delta d/d$ resolution, simple calculations show that the number of resolved peaks, N_R, in a time-of-flight powder diffraction pattern is of the order of

$$N_R = 1/[3(\Delta d/d)]. \tag{11.22}$$

Allowing a factor of $\sim 7{:}1$ for the ratio of the number of resolved peaks to structural parameters refined to high precision, implies that for $(\Delta d/d) \sim 5 \times 10^{-4}$ (as obtained with the High Resolution Powder Diffractometer HRPD at ISIS) the number of potential structural parameters in a high-precision study is approximately 100. As a test of these assumptions, the crystal structure of deuterated benzene was investigated with the HRPD (ISIS 1989). The recent single-crystal neutron structure refinement by Jeffrey et al. (1987) and theoretical calculations by Filippini and Gramaccioli (1989) provided rigorous tests for the powder data collected on HRPD.

Benzene, C_6H_6, is one of the most important organic molecules, forming the basic building unit of all aromatic compounds. Because of its central role in organic chemistry, benzene has been extensively studied by numerous experimental and theoretical techniques. The simplicity of the chemical formula, C_6H_6, belies, however, the complexity of its crystal structure. As a result of a complex packing configuration, benzene adopts an orthorhombic structure, space group $Pbca$ ($p = 2$), with a moderately-sized unit cell ($a = 7.3550$ Å, $b = 9.3709$ Å, $c = 6.6992$ Å, $V = 461.7$ Å^3).

Benzene melts at 6 °C. A powder sample was, therefore, prepared by grinding 5 cm^3 of deuterated benzene in a glove box under a cold nitrogen atmosphere. The sample was then loaded into a cylindrical vanadium can and rapidly cooled to liquid helium temperatures (4 K) to avoid problems with preferred orientation. Data were collected at the high resolution 2 m position ($\Delta d/d \sim 5 \times 10^{-4}$) on HRPD for approximately 9 h (174 µA h). The raw data were corrected for incident flux (using a vanadium calibration), and cryostat and sample attenuation. The last correction was derived from a measurement of the transmitted neutron flux and showed significant structure from multiple-scattering self-attenuation effects.

Rietveld refinement was performed with the powder diffraction analysis package developed at RAL (David *et al.* 1988) based upon the Cambridge Crystallography Subroutine Library. The data ranged in *d*-spacing from 0.606 Å to 1.778 Å, consisted of 5382 points, and included 1040 reflections. Small impurity peaks were present in the diffraction pattern. They were caused, somewhat remarkably, by the vanadium sample can and cryostat heat shields, even though vanadium has a near-zero coherent neutron cross-section. At present, these peaks have not been considered in the data analysis. With the use of a peak shape consisting of a double-exponential decay convoluted with a Voigt function (itself the convolution of Gaussian and a Lorentzian function), an excellent least-squares fit to the powder diffraction data was obtained, as is shown in Fig. 11.10. Perhaps most remarkable about this Rietveld refinement profile is the intricate detail of the data at small *d*-spacing arising from the unusually high resolution of the HRPD. These data provide a graphic example of the large amount of information that can be obtained from powder diffraction data when a high-resolution diffractometer is employed and when the sample itself exhibits a narrow intrinsic Bragg peak width.

The refined structural parameters, including 18 atomic coordinates and 36 anisotropic temperature factors, are listed in Table 11.3. Table 11.3 also lists bond lengths, uncorrected for libration, obtained in the present study and from the work of Jeffrey *et al.* (1987). The agreement is good. The few statistically significant differences probably result from systematic errors in the powder diffraction data. Experimental and theoretically calculated anisotropic temperature factors are presented in Table 11.4. With the exception of the B_{22} temperature factors for three carbon atoms, there is a remarkable agreement between the temperature factors obtained from the HRPD powder data and from the single-crystal data.

The results obtained from Rietveld refinement with HRPD data are only marginally inferior to the best single crystal data. The two experimental techniques agree closely with each other, and both differ from the theoretical calculations (Fillipini and Gramaccioli 1989), particularly in the values

---

Fig. 11.10 (*opposite*)   The final Rietveld refinement profile for data from benzene obtained on the HRPD at ISIS. The first four frames (a)–(d) show the fit for *d*-spacings from 0.5 to 2.0 Å. The final two frames (e)–(f) show the fit for the range 0.68–0.78 Å on an expanded scale in order to better illustrate the complexity of the data and the quality of the fit at small *d*-spacings. Tick marks at the top of each plot indicate the positions of the Bragg reflections. A difference curve (observed minus calculated) plotted in units of the statistical uncertainty, σ (square root of the number of counts) is plotted at the bottom. The dotted lines in the difference curves are at $+/-3\sigma$.

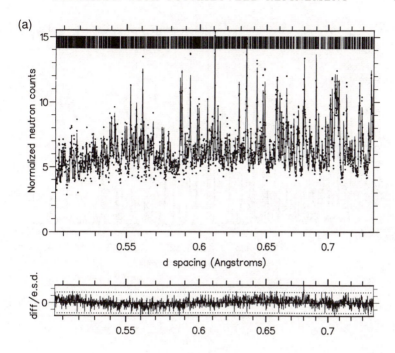

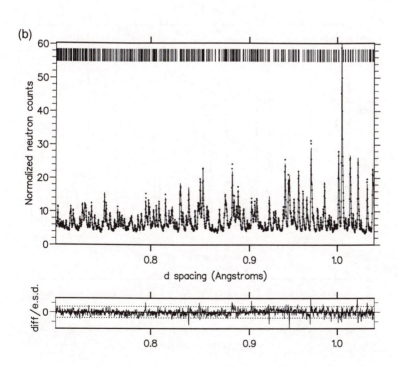

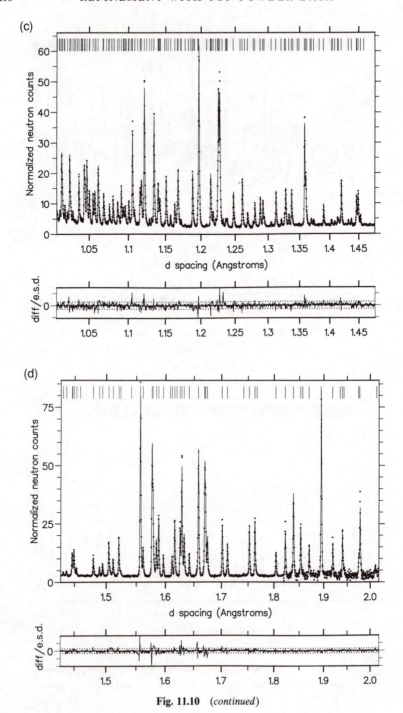

Fig. 11.10 (*continued*)

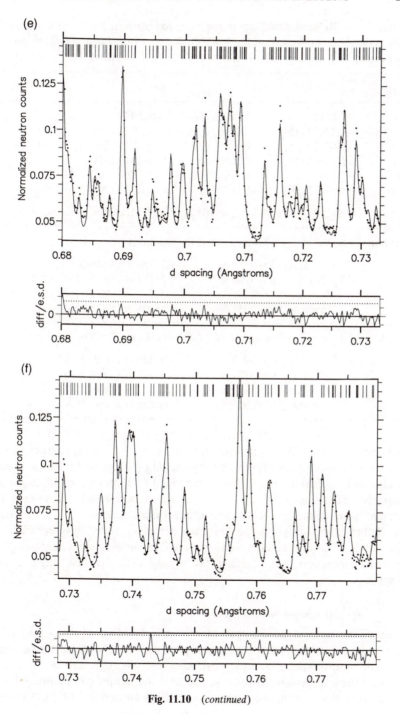

Fig. 11.10 (*continued*)

Table 11.3 Refined structural parameters for benzene from neutron powder diffraction data collected on the HRPD. Space group $Pbca$, $p = 2$, molecular symmetry $= \bar{1}$, $a = 7.3551(3)$ Å, $b = 9.3712(4)$, $c = 6.6994$ Å, $V = 461.76$ Å^3

Atom	x/a	y/b	z/c	B_{iso} ($\times 10^{-4}$ Å^2)
C1	−0.06120 (15)	0.14123 (10)	−0.00519 (20)	68 (6)
C2	−0.14023 (15)	0.04469 (10)	0.12722 (15)	66 (6)
C3	−0.07770 (15)	−0.09689 (12)	0.13264 (20)	77 (6)
D1	−0.10853 (15)	0.25050 (15)	−0.01187 (25)	202 (9)
D2	−0.24908 (20)	0.07682 (15)	0.22600 (20)	202 (9)
D3	−0.13821 (20)	−0.17136 (15)	0.23703 (20)	203 (9)

$R_p = 13.2\%$, $R_{wp} = 15.7\%$, $R_{exp} = 10.8\%$, $\chi^2 = 2.1$ (usual R-factor notation)

Benzene: bond-lengths (4 K) (uncorrected for libration)

C1–C2 = 1.3940(20) Å	C1–D1 = 1.0825(30) Å
C2–C3 = 1.4047(30) Å	C2–D2 = 1.0815(30) Å
C1–C3 = 1.3948(20) Å	C3–D3 = 1.0836(25) Å
mean = 1.3978(15) Å	mean = 1.0825(20) Å

Single crystal neutron diffraction (15 K) (after Jeffrey *et al.* 1987)

C1–C2 = 1.3969(7) Å	C1–D1 = 1.0879(9) Å
C2–C3 = 1.3970(8) Å	C2–D2 = 1.0869(9) Å
C1–C3 = 1.3976(7) Å	C3–D3 = 1.0843(8) Å
mean = 1.3972(5) Å	mean = 1.0864(7) Å

obtained for the anisotropic temperature factors for the deuterium atoms. The powder diffraction experiment thus strongly supports the single crystal study and indicates that further improved theoretical calculations are required. The quality of these powder diffraction results represents the present state-of-the-art at ISIS. Further improvements in normalization procedure and multiphase analysis are currently under development and should lead to a precision and accuracy in moderately complex structure determination that compare favourably with the best single crystal results.

11.10 Special sample environments

One of the natural advantages of the pulsed-source neutron diffraction technique is the ability to provide complete data at a fixed scattering angle. For the case of special sample environments, the angle can be chosen to provide the best possible collimation so that the sample can be probed with no unwanted scattering from the surroundings (Jorgensen 1988). Although

Table 11.4 Anisotropic temperature factors ($\times 10^4$ Å^2) for benzene at low temperatures

Atom	B_{11}	B_{22}	B_{33}	B_{23}	B_{13}	B_{12}
C1	90	66	89	3	7	0
	77	58	77	3	0	7
	79 (2)	67 (2)	88 (2)	4 (1)	7 (2)	6 (2)
	77 (7)	42 (6)	87 (7)	1 (5)	5 (5)	−3 (4)
C2	84	87	82	−2	17	6
	71	79	70	−2	17	9
	74 (2)	81 (2)	79 (2)	0 (2)	17 (2)	9 (2)
	71 (7)	58 (7)	68 (6)	9 (4)	26 (5)	12 (4)
C3	86	79	82	10	11	6
	73	72	70	10	11	−5
	81 (2)	75 (2)	82 (2)	10 (1)	14 (2)	−3 (2)
	83 (7)	57 (7)	92 (7)	0 (5)	18 (5)	−1 (4)
D1	216	173	212	11	39	38
	222	165	199	11	39	38
	224 (3)	114 (2)	239 (3)	12 (2)	25 (2)	46 (2)
	218 (8)	121 (7)	267 (9)	19 (5)	22 (6)	31 (5)
D2	184	226	215	5	56	56
	170	217	202	5	55	56
	183 (2)	204 (3)	208 (3)	−8 (2)	88 (2)	35 (2)
	170 (8)	212 (8)	225 (9)	−2 (6)	120 (6)	33 (5)
D3	228	208	171	60	37	−2
	215	199	158	59	36	−1
	214 (3)	171 (2)	199 (3)	58 (2)	61 (2)	−18 (4)
	241 (9)	155 (8)	214 (8)	75 (6)	66 (7)	−20 (5)

there are advantages for all types of special sample environments, the most important cases are those in which the sample containment vessel must be close to, or in contact with, the sample.

The clearest example of such a case is diffraction at high pressure, where the pressure cell, in fact, must support the sample. As has already been mentioned, diffraction studies at high pressure were the first widespread applications of the time-of-flight technique, using chopper-based instruments at reactor sources. This work has continued at the pulsed neutron sources and, of course, has benefited from the improvements in both resolution and flux. Jorgensen has recently reviewed structural studies at high pressure using time-of-flight neutron powder diffraction (Jorgensen 1990). The most important advance resulting from the high flux and resolution at the pulsed sources has been the ability to solve and then refine unknown structures from the powder diffraction data. For example, the structure of KNO_3-IV was solved by indexing the observed d-spacings to obtain the unit cell, determining the space group from the observed extinctions, and then refining by the Rietveld method for a trial structure based on analogy to other structures in the literature (Worlton et al. 1986). The most complex structure solved to date from neutron powder diffraction data taken at high pressure is that of ND_4F-II (Lawson et al. 1989). The structure contains 24 molecules in a 1024 $Å^3$ hexagonal unit cell.

More recently, many of the newly discovered oxide superconductors have been studied by time-of-flight neutron powder diffraction as a function of pressure. Some of these compounds exhibit the largest pressure dependence of the superconducting transition temperature, dT_c/dP, observed for any superconductors (Murayama 1989). The goal of neutron diffraction measurements at high pressure is to determine whether pressure-induced changes in particular structural features can be correlated with the changes in T_c. Such information provides an important test of models proposed to explain the behaviour of these compounds in terms of the distribution of charge. The bond lengths determined from Rietveld refinement are used to determine the charge distribution. Fig. 11.11 shows the Rietveld refinement profile of data for $YBa_2Cu_3O_{6.93}$ at a pressure of 0.58 GPa (Jorgensen et al. 1990b). The data were collected on the Special Environment Powder Diffractometer at Argonne's IPNS with the sample in a helium-gas pressure cell specially designed for time-of-flight diffraction. One of the most striking features of these data is the complete absence of any scattering from the pressure cell. This is achieved by incorporating neutron shielding embedded in the walls of the pressure cell to eliminate all scattered neutrons except those scattered from the sample at $2\theta = 85°$–$95°$ (Jorgensen et al. 1990b). Using this technique, it is possible to achieve Rietveld refinements for samples in high pressure environments of a quality equal to that achieved under ambient conditions.

Another area in which the time-of-flight technique has had major impact is for high temperature diffraction. Here, for achieving high temperatures with small thermal gradients it is advantageous to have the walls of the furnace fairly close to the sample. Again, the ability to collect data at a fixed scattering angle (typically $2\theta = 90°$) is a clear advantage. Perhaps the most famous high temperature neutron diffraction experiments in recent years were the studies of oxide superconductors in which oxygen site occupancies varied over a wide range. In the case of $YBa_2Cu_3O_{7-x}$ these experiments must be done in atmospheres with controlled oxygen partial pressures (Jorgensen et al. 1987; Jorgensen et al. 1989b). From the data, the site occupancies of particular oxygen sites were obtained by Rietveld refinement,

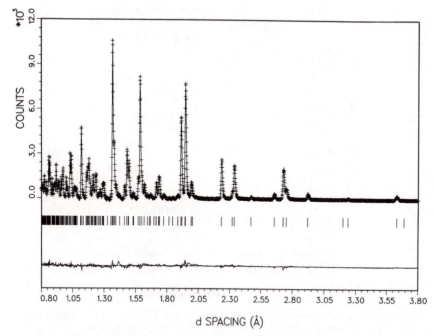

Fig. 11.11 Rietveld refinement plot for $YBa_2Cu_3O_{6.93}$ at 0.58 GPa in a helium-gas pressure cell. Scattering from the aluminum pressure cell is completely eliminated by incorporating neutron shielding in the walls of the cell in order to restrict the scattered beam to neutrons scattered from the sample at $2\theta = 85°-95°$. (From Jorgensen *et al.* 1990*b*).

as is shown in Fig. 11.12. These high temperature *in situ* measurements were then correlated with the superconducting properties for oxygen-deficient $YBa_2Cu_3O_{7-x}$, which vary as a function of both the oxygen concentration and the ordering of oxygen atoms on the available sites (Jorgensen *et al.* 1990).

11.11 Future developments

Although remarkable advances in time-of-flight powder diffraction methods at pulsed neutron sources have occurred in recent years, these methods may be properly viewed as being in their infancy. The ultimate flux capabilities of the spallation pulsed neutron sources have not yet been reached. Based on presently known design concepts, one to two additional orders of magnitude in flux appear feasible (Lander and Emery 1985). Clearly this will allow the design of yet another generation of sophisticated instrumentation. Many new instrument design concepts are already known, and are awaiting the opportunity to be tested. These future developments will undoubtedly include the extension of experimental capabilities to higher

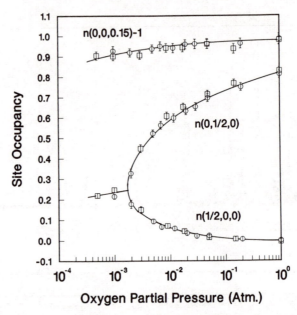

Fig. 11.12 Site occupancies of three of the oxygen sites in $YBa_2Cu_3O_{7-x}$ vs oxygen partial pressure at a constant temperature of 490 °C. The site occupancies are obtained from Rietveld refinement of neutron powder diffraction data taken *in situ* in a flowing atmosphere. An orthorhombic-to-tetragonal transition occurs where the occupancies of the $(0, 1/2, 0)$ and $(1/2, 0, 0)$ sites meet. In the tetragonal phase these two sites are symmetry equivalent. (From Jorgensen *et al.* 1989*b*.)

resolution, smaller samples, and more rapid data collection for real time measurements. At the same time, it is clear that the general concepts of Rietveld refinement, i.e. the concept that a calculated model of the diffraction data can be compared directly with the complete diffraction 'profile', will be expanded to allow access to the new information available in the data. In a single refinement with such data, one could expect to obtain not only the crystal structure parameters, but also such information as the parameters describing commensurate or incommensurate structural modulations (see, for example, Chapter 13 by Izumi), the magnitude and direction of macro-strain and microstrain, parameters describing the texture, and the configurations of point and extended defects including the local pair distributions that give rise to diffuse background scattering.

11.12 Acknowledgements

W. I. F. David is supported by the UK Science and Engineering Research Council. J. D. Jorgensen is supported by the US Department of Energy, Office of Basic Energy Science, Division of Materials Sciences, under contract number W-31-109-ENG-38.

References

Albinati, A. and Willis, B. T. M. (1982). *J. Appl. Crystallogr.*, **15**, 361.

Brown, B. S., Carpenter, J. M., Jorgensen, J. D., Price, D. L., and Kamitakahara, W. (1982). In *Novel materials and techniques in condensed matter* (ed. G. W. Crabtree and P. Vashishta), pp. 311–40. North Holland, New York.

Buras, B. and Gerward, L. (1975). *Acta Crystallogr.*, **A31**, 372.

Carpenter, J. M. (1967). *Nucl. Instrum. Methods*, **47**, 179.

Carpenter, J. M. (1977). *Nucl. Instrum. Methods*, **145**, 91.

Carpenter, J. M., Mueller, M. H., Beyerlein, R. A., Worlton, T. G., Jorgensen, J. D., Brun, T. O. *et al.* (1975). In *Proc. Neutron Diffraction Conf.*, Petten, The Netherlands, 5–6 August 1975 (Reactor Centrum Nederland, RCN-234), pp. 192–208.

Carpenter, J. M., Price, D. L., and Swanson, N. J. (1978). Argonne National Laboratory Report No. ANL-78-88.

Cartz, L. and Jorgensen, J. D. (1981). *J. Appl. Phys.*, **52**, 236.

Cartz, L., Srinivasa, S. R., Riedner, R. J., Jorgensen, J. D., and Worlton, T. G. (1979). *J. Chem. Phys.*, **71**, 1718.

Champeney, D. C. (1973). *Fourier transforms and their physical applications*. Academic Press, London.

Cole, I. and Windsor, C. G. (1980). *Nucl. Instrum. Methods*, **171**, 107.

David, W. I. F. (1990). In *Neutron scattering for materials science*, Materials Research Society Symposium Proceedings, Vol. 156, Pittsburgh (ed. S. M. Shapiro, S. C. Moss, and J. D. Jorgensen), pp. 41–54.

David, W. I. F., Akporiaye, D. E., Ibberson, R. M., and Wilson, C. C. (1988). Rutherford Appleton Laboratory Report No. RAL-88-103.

David, W. I. F., Moze, O., Licci, F., Bolzoni, F., Cywinski, R., and Kilcoyne, S. (1989). *Physica*, **B156 & 157**, 884.

Decker, D. L., Beyerlein, R. A., Roult, G., and Worlton, T. G. (1974). *Phys. Rev. B*, **10**, 3584.

Fillipini, G. and Gramaccioli, C. M. (1989). *Acta Crystallogr.*, **A45**, 261.

Graham, K. F. and Carpenter, J. M. (1972). *Nucl. Sci. Eng.*, **49**, 418.

Ikeda, S. and Carpenter, J. M. (1985). *Nucl. Instrum. Methods*, **A239**, 536.

ISIS Annual Report (1989). Rutherford Appleton Laboratory.

Izumi, F., Asano, H., Murata, H., and Watanabe, N. (1982). *J. Appl. Crystallogr.*, **20**, 581.

Jeffrey, G. A., Ruble, J. R., McMullan, R. K., and Pople, J. A. (1987). *Proc. Roy. Soc. Lond.*, **A414**, 47.

Jorgensen, J. D. (1978). *J. Appl. Phys.*, **49**, 5473.

Jorgensen, J. D. (1988). In *Chemical crystallography with pulsed neutrons and synchrotron X-rays*, NATO ASI Series C, Mathematical and Physical Sciences, Vol. 221 (ed. M. A. Carrondo and G. A. Jeffrey), pp. 159–86. D. Reidel, Dordrecht.

Jorgensen, J. D. (1990). In *High pressure science and technology*, Proc. 12th AIRAPT and 27th EHPRG International Conference, Paderborn, Germany, 17–21 July 1989 (ed. W. B. Holzaphel and P. G. Johansen), pp. 441–3. Gordon and Breech, London. Also published as (1990). *High Pressure Res.*, **4**, 441.

Jorgensen, J. D. and Clark, J. B. (1980). *Phys. Rev.*, **B22**, 6149.

Jorgensen, J. D. and Rotella, F. J. (1982). *J. Appl. Crystallogr.*, **15**, 27.

Jorgensen, J. D., Worlton, T. G., and Jamieson, J. C. (1978a). *Phys. Rev.*, **B17**, 2212.

Jorgensen, J. D., Johnson, D. H., Mueller, M. H., Peterson, S. W., Worlton, T. G., and Von Dreele, R. B. (1978b). In *Proc. Conf. on Diffraction Profile Analysis*, Cracow, Poland, 14–15 August 1978, pp. 20–2.

Jorgensen, J. D., Beyerlein, R. A., Watanabe, N., and Worlton, T. G. (1984). *J. Chem. Phys.*, **81**, 3211.

Jorgensen, J. D., Beno, M. A., Hinks, D. G., Soderholm, L., Volin, K. J., Hitterman, R. L. *et al.* (1987). *Phys. Rev.*, **B36**, 3608.

Jorgensen, J. D., Faber, J., jr, Carpenter, J. M., Crawford, R. K., Haumann, J. R., Hitterman, R. L. *et al.* (1989a). *J. Appl. Crystallogr.*, **22**, 321.

Jorgensen, J. D., Shaked, H., Hinks, D. G., Dabrowski, B., Veal, B. W., Paulikas, A. P. *et al.* (1989b). *Physica*, **C153–155**, 578.

Jorgensen, J. D., Veal, B. W., Paulikas, A. P., Nowicki, L. J., Crabtree, G. W., Claus, H., and Kwok, W. K. (1990a). *Phys. Rev.*, **B41**, 1863.

Jorgensen, J. D., Pei Shiyou, Lightfoot, P., Hinks, D. G., Veal, B. W., Dabrowski, B. (1990b). *Physica*, **C**, **171**, 93.

Lander, G. H. and Emergy, V. J. (1985). *Nucl. Instrum. Methods Phys. Res.*, **B12**, 525.

Lawson, A. C., Roof, R. B., Jorgensen, J. D., Morosin, B., and Schirber, J. E. (1989). *Acta Crystallogr.*, **B45**, 212.

Mueller, M. H., Beyerlein, R. A., Jorgensen, J. D., Brun, T. O., Satterthwaite, C. B., and Caton, R. (1977). *J. Appl. Crystallogr.*, **10**, 79.

Murayama, C., Mori, N., Yomo, S., Takagi, H., Uchida, S., and Tokura, Y. (1989). *Nature*, **339**, 293.

Rietveld, H. M. (1969). *J. Appl. Crystallogr.*, **2**, 65.

Sabine, T. M. (1988). *Acta Crystallogr.*, **A44**, 368.

Sabine, T. M., von Dreele, R. B., and Jorgensen, J.-E. (1988). *Acta Crystallogr.*, **A44**, 374.

Steichele, E. and Arnold, P. (1973). *Phys. Lett.*, **A44**, 165.

von Dreele, R. B., Jorgensen, J. D., and Windsor, C. G. (1982). *J. Appl. Crystallogr.*, **15**, 581.

Windsor, C. G. (1981). *Pulsed neutron scattering*. Wiley, New York.

Windsor, C. G. and Sinclair, R. N. (1976). *Acta Crystallogr.*, **A32**, 395.

Windsor, C. G., Bunce, L. J., Borcherds, P. H., Cole, I., Fitzmaurice, M., Johnson, D. A. G., and Sinclair, R. N. (1980). *Nucl. Instrum. Methods*, **171**, 107.

Worlton, T. G. and Beyerlein, R. A. (1975). *Phys. Rev.*, **B12**, 1899.

Worlton, T. G., Jorgensen, J. D., Beyerlein, R. A., and Decker, D. L. (1976). *Nucl. Instrum. Methods*, **137**, 331.

Worlton, T. G., Decker, D. L., Jorgensen, J. D., and Kleb, R. (1986). *Physica*, **B136**, 503.

12

Combined X-ray and neutron Rietveld refinement

Robert B. Von Dreele

12.1 Introduction

As is noted in Chapter 1, a neutron or X-ray powder diffraction pattern can be modelled from descriptions of the crystal structure of the sample, both instrumental and sample contributions to peak broadening, intensity effects such as absorption and extinction, and a background contribution. Also noted is that a Rietveld (1969) refinement performs a least-squares minimization of the weighted differences between this calculated pattern and observed data by computing the shifts in the adjustable parameters for the model. Since the problem is non-linear, several cycles of least-squares refinement are needed to complete the minimization. However, because both the complex nature of the structure factors and much of the directional character of reciprocal space are lost in a powder diffraction experiment, there is a larger question of uniqueness for the result of a Rietveld refinement of a single powder pattern than for the result of a single crystal refinement.

A possible solution to this problem is suggested by the fact that X-rays and neutrons are scattered by different mechanisms. X-ray scattering occurs primarily by interaction with the electrons that surround an atom, so that elements at the top of the periodic table are weaker X-ray scatterers than those near the bottom and the scattering power falls off with increasing angle. Thus, the 'normal' scattering factors for X-rays look like those shown in Fig. 12.1. If the wavelength is close to an absorption edge, the scattering factor curves can be greatly modified by anomalous dispersion to give curves such as those pictured in Fig. 12.2. On the other hand, neutrons are primarily scattered by atomic nuclei so there is only isotropic point scattering. Also, nuclear scattering is a combination of scattering by short range nuclear forces and resonance scattering which gives neutron scattering lengths that vary

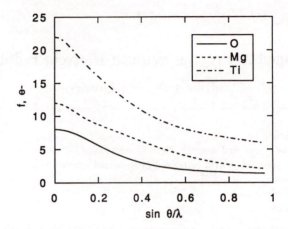

Fig. 12.1 Normal X-ray scattering factor curves for Ti, Mg, and O as a function of $\sin \theta/\lambda$.

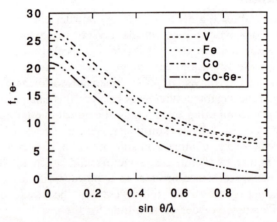

Fig. 12.2 Normal X-ray scattering factors for V, Fe, and Co; that for radiation 6 eV below the K-edge for Co is also shown.

very erratically across the periodic table and from isotope to isotope (Figure 12.3). Thus, X-ray and neutron powder diffraction patterns will be very different. Figure 11.4 shows idealized X-ray and neutron patterns calculated for $MgTiO_3$ 'geikielite'. These were generated for essentially identical diffractometer experiments (impossible in real life) but are startingly different. In fact, the strongest peak in the X-ray pattern (at $\sim 32°$) is completely absent in the neutron pattern!

The reason for the extreme difference between these two patterns lies in the atomic scattering factors for titanium, magnesium, and oxygen for X-rays and neutrons. The X-ray scattering factors are simply proportional to the

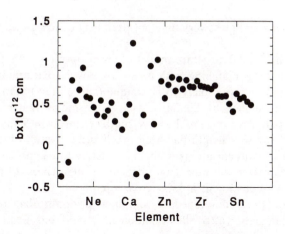

Fig. 12.3 Neutron scattering lengths for the natural abundance element H–Xe.

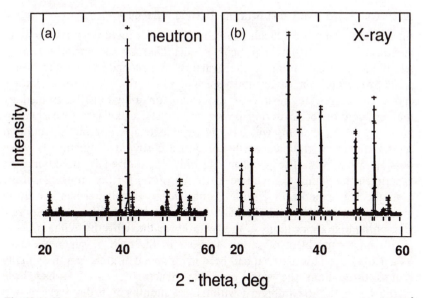

Fig. 12.4 Simulated powder diffraction patterns for $MgTiO_3$ geikielite for $\lambda = 1.5428$ Å radiation; (a) neutron pattern and (b) X-ray pattern.

atomic number, thus $f_{Ti} > f_{Mg} > f_O$. However, the titanium scattering length for neutrons is negative and the value for oxygen is only slightly larger than magnesium, thus $b_O > n_{Mg} > 0 > b_{Ti}$. The consequence is that for each reflection, the neutron structure factor is very different from the X-ray structure factor and gives very different peak heights in a powder diffraction pattern.

Clearly, the crystal structure model of atom positions, etc. that leads to calculated patterns matching both a neutron powder pattern and an X-ray powder pattern is more likely to be unique (and correct) than that from a single pattern refinement. Similarly, a structure model that matches all of the powder patterns from a multi-wavelength synchrotron X-ray experiment, where the X-ray wavelength has been tuned successively to the absorption edges of the constituent atoms is also more likely to be correct because the model has to match the wavelength-dependent differences in the scattering factors for the constituent atoms. To accommodate this idea, Larson and Von Dreele (1987) developed a computer program that performs the necessary calculations to allow combined X-ray and neutron Rietveld refinement of a crystal structure. The remainder of this chapter is a discussion of two applications of this idea.

12.2   Combined X-ray and neutron Rietveld refinement

Intrinsically, the most difficult problem in crystal structure analysis is atom identification. An atom is principally identified by its scattering power relative to the other atoms in the structure, and secondary identification is made by crystal chemical arguments, i.e. bond length comparison to known crystal structures. In most case, atoms are unambiguously identified by the combination of scattering power and crystal chemistry. Occasionally, however, different atoms will have nearly identical scattering powers and crystal chemistry arguments will not give a clear cut distinction between possible atom sites. The problem can only be resolved by modifying the experiment to change the relative scattering power for the atoms; generally a single new experiment is still insufficient to clearly resolve the atom identification problem and an analysis of the combined data set is required. One of the first applications of this method was the work by Williams *et al.* (1988) on the problem of cation disorder in the high T_c superconductor $YBa_2Cu_3O_{7-x}$. This material had been investigated at great length by many groups throughout the world and the structure (Fig. 12.5) had been established with little ambiguity within a few months of its discovery by Wu *et al.* (1987). Almost all of these structural studies were the result of Rietveld refinements using neutron powder diffraction data obtained at either reactor or pulsed spallation sources. By unfortunate coincidence the neutron scattering lengths of yttrium and copper are virtually identical, so it was possible that the assignment of these two atom locations was in fact reversed. An interpretation of the EXAFS of this material (Lytle and Greegor 1988) suggested the possibility that the Y atom site is 1/3 Cu, contrary to crystal chemical expectations from interatomic distances and coordination numbers.

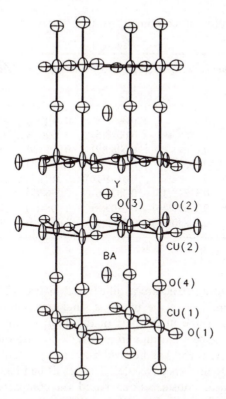

Fig. 12.5 Structure of orthorhombic $YBa_2Cu_3O_{7-x}$ obtained from combined Rietveld refinement of TOF neutron and Cu K_α powder diffraction data. The ellipsoids correspond to 90% probability surfaces for the atomic thermal motion.

However, the X-ray scattering factors for these two atoms are very different and a refinement combining both X-ray and neutron diffraction data would remove the ambiguity.

The Rietveld refinement of this structure consisted of fitting four neutron time-of-flight patterns taken on the High Intensity Powder Diffractometer (HIPD) at LANSCE and two X-ray powder patterns collected on a commercial instrument with Cu K_α radiation. The entire set of six powder patterns contained $\sim 25\,000$ data points and were fitted with ~ 120 adjustable parameters. These parameters included the 33 needed to describe the crystal structure of $YBa_2Cu_3O_{7-x}$, namely, atomic positions, fractional occupancies, thermal parameters, and lattice parameters. The rest characterized details of the powder diffraction patterns. These included coefficients for the background functions needed for each powder pattern, coefficients for the powder diffraction peak shapes, and intensity correction coefficients for absorption, preferred orientation, and extinction as well as six scaling

Table 12.1 Results of combined neutron/X-ray Rietveld refinement of $YBa_2Cu_3O_{7-x}$

Space group, *Pmmm*, $a = 3.82030(8)$, $b = 3.88548(10)$, $c = 11.68349(23)$ Å

Atom	x	y	z	U_{11}	U_{22}	U_{33}	Fraction
Y	1/2	1/2	1/2	0.0085(8)	0.0106(8)	0.0085(6)	1
Ba	1/2	1/2	0.18393(6)	0.0078(6)	0.0096(7)	0.0198(5)	1
Cu(1)	0	0	0	0.0080(9)	0.0115(9)	0.0150(7)	1
Cu(2)	0	0	0.35501(8)	0.033(5)	0.0036(5)	0.0207(5)	1
O(1)	0	1/2	0	0.0161(16)	0.0104(11)	0.0080(14)	0.910(8)
O(2)	1/2	0	0.37819(15)	0.0039(6)	0.0068(7)	0.0203(11)	1
O(3)	0	1/2	0.37693(16)	0.0109(8)	0.0084(7)	0.0056(11)	1
O(4)	0	0	0.15840(13)	0.0162(11)	0.0123(9)	0.0097(7)	1

factors. The resulting structure as illustrated in Fig. 12.5 was dramatically more precise than any of the previous single measurement results (Table 12.1). Some idea of the fit for part of the data is seen in Fig. 12.6 which shows the same *d*-spacing range from one X-ray and one neutron pattern. Most importantly, there was no evidence of any interchange between the metal atoms on their respective sites. This result had been expected because of crystal chemical considerations based on comparison of interatomic distances and ionic radii. In addition, this refinement gave a complete description of the thermal motion for all the atoms which was different from the previous studies and, because of the requirement to fit both X-ray and neutron diffraction data, is probably more accurate as well.

12.3   Rietveld refinement with combined neutron and synchrotron X-ray data

In a recent study by Williams *et al.* (1988), the problem of metal atom ordering in the commercial ferromagnetic material, FeCo (2 per cent V), was addressed by combining neutron TOF powder diffraction data with synchrotron X-ray data including some with wavelengths tuned near the K absorption edges of the constituent atoms. Since the pure alloy FeCo is a difficult material to use in industrial applications because of its brittleness, a small amount (2–5 per cent) of vanadium is usually added to improve its machinability. It had generally been assumed that the V occupied equally both metal sites in this BCC metal, but no evidence was available to confirm that notion. Mal'tsev *et al.* (1975) suggested from the magnetic behaviour of dilute alloys that V should prefer the Fe sublattice, and recent Mössbauer studies by Persiano and Rawlings (1987) suggest that assignment is true.

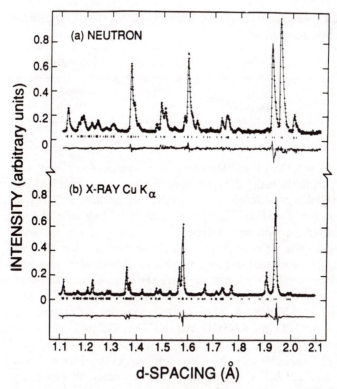

Fig. 12.6 Part of the X-ray and neutron diffraction data for orthorhombic $YBa_2Cu_3O_{7-x}$. Data shown by plus marks in (a) represent data collected on the $+153°$ 2θ bank of the HIPD diffractometer at LANSCE and (b) represent data taken on a Siemens D500 X-ray diffractometer with Cu K_α radiation. The solid lines are the best fit to the data from the combined refinement. Tick marks show positions for allowed reflections; those for Cu $K_{\alpha2}$ are shown shifted from the d-spacings calculated for Cu $K_{\alpha1}$. The lower curve in each panel represents the difference between the observed and calculated profiles.

The alloy exhibits a classic β-brass-type order–disorder phase transition at 720 °C from a partially ordered CsCl structure to a true BCC structure. Previous studies on the binary alloy by neutron diffraction (Mal'tsev *et al.* 1975; Smith and Rawlings 1976; Glazyrina *et al.* 1983) and Co K_α X-ray diffraction (Buckley 1975; Clegg and Buckley 1975) have established the ordering behaviour of Fe and Co. However, determination of the ordering scheme for all three atoms over the two sites in the V doped material is not possible from any single diffraction study, only a multiple radiation experiment could solve the problem.

The scattering lengths of Fe, Co, and V are 0.954, 0.253, and -0.0038×10^{-12} cm, respectively, so these atoms are relatively easy to

Table 12.2   Results of combined neutron/synchrotron X-ray Rietveld refinement of FeCo(2% V) alloy

Site	Fraction Fe	Fraction Co	Fraction V	$U(\text{Å}^2)$
0, 0, 0	0.852(3)	0.169(8)	−0.001(7)	0.0042(1)
1/2, 1/2, 1/2	0.148(3)	0.791(8)	0.041(7)	0.0026(2)

distinguish with neutrons. However, the normal X-ray scattering factors, Fig. 12.2, are in the ratio of their atomic numbers (26, 27, and 23, respectively) which makes them difficult to distinguish. By tuning the X-ray wavelength to be just below the K absorption edge by ~5 eV, the real part of the anomalous dispersion can reduce the scattering factor by ~6e− for all scattering angles. This changes both the magnitude and the shape of the scattering factor for that one element (Fig. 12.2) making it easy to distinguish it from other similar elements. To solve the three atom occupancy problem for the FeCo(V) ferromagnet, five diffraction experiments were performed (Williams *et al.* 1990). A TOF neutron powder diffraction pattern was obtained on HIPD at LANSCE, and the X-ray diffraction experiments were done on the powder diffractometer on line IV-3 at SSRL. The four wavelengths were chosen to be just below each of the three K-edges (1.74343, 1.60838, and 2.26965 Å, respectively) and one much shorter (1.23917 Å) to get a wide *d*-spacing range free of any anomalous dispersion effects. The structure was refined by multiple data set Rietveld refinement with GSAS (Larson and Von Dreele 1987) using 22 individual powder pattern scans consisting of 4 neutron TOF patterns and 18 X-ray patterns with 4 wavelengths. The structural parameters consisted of 2 temperature factors for the two atom sites and 3 occupation factors for the constituent elements. Because the chemical analysis was known, the total occupation factor for each element was fixed. The results (Table 12.2) showed a strong preference of V for the Co rich site, in contrast with the Mössbauer result. The sums of the site fractions are very close to unity (1.02 for the Fe rich site and 0.98 for the Co rich site) which provide a useful consistency check on the result.

12.4   Conclusions

As the examples demonstrate, appropriate combination of different powder diffraction data sets in a single Rietveld refinement can solve structural problems that cannot be addressed with a single data set. For the difficult problem of atom site distributions, the combinations should be chosen to enhance the contrast between the different atoms.

12.5 Acknowledgements

The authors wishes to acknowledge the support of the Office of Basic Energy Sciences, US Department of Energy for this work and the collaboration of Allen C. Larson during the development of the software GSAS.

References

Buckley, R. A. (1975). *Metal Sci. J.*, **9**, 243.

Clegg, D. W. and Buckley, R. A. (1975). *Metal Sci. J.*, **7**, 48.

Glazyrina, M. I., Glezer, A. M., Molitilov, B. V., Tret'yakova, S. M., and Kleynerman, V. I. (1983). *Fiz. met. metall.*, **56**, 733.

Larson, A. C. and Von Dreele, R. B. (1987). Los Alamos National Laboratory Report No. LA-UR-86-748.

Lytle, F. W. and Greggor, R. B. (1988). *Phys. Rev.*, **B37**, 1550.

Mal'tsev, Ye. I., Goman'kov, V. I., Monkov, B. N., Puzey, I. M., and Nogin, N. I. (1975). *Fiz. met. metall.*, **40**, 443.

Persiano, A. I. C. and Rawlings, R. D. (1987). *Phys. Status Solidi*, **A103**, 547.

Rietveld, H. M. (1969). *J. Appl. Crystallogr.*, **2**, 65.

Smith, A. W. and Rawlings, R. D. (1976). *Phys. Status Solidi*, **A34**, 117.

Williams, A., Kwei, G. H., Von Dreele, R. B., Larson, A. C., Raistrick, I. D., and Bish, D. L. (1988). *Phys. Rev.*, **B37**, 7960.

Williams, A., Kwei, G. K., Ortiz, A. T., Karnowski, M., and Warburton, W. K. (1990). *J. Mater. Sci.*, **6**, 1197.

Wu, M. K., Ashburn, J. R., Torng, C. J., Hor, P. H., Meng, R. L., Gao, L. *et al.* (1987). *Phys. Rev. Lett.*, **58**, 908.

13

Rietveld analysis programs RIETAN and PREMOS and special applications

F. Izumi

13.1 Description

13.1.1 Main features

A FORTRAN program named RIETAN (**RIET**veld **AN**alysis) was independently developed in Japan for Rietveld analysis of angle-dispersive X-ray and neutron powder data (Izumi 1985; Izumi 1989). It has since become the heart of a suite of related programs called FAT–RIETAN (Section 13.2). RIETAN was later modified to work with TOF neutron powder data (Izumi et al. 1987) measured on the high-resolution powder diffractometer, HRP (Watanabe et al. 1987), installed at the KENS pulsed neutron source at the National Laboratory for High Energy Physics (KEK). Since 1983, structures of many metals and inorganic compounds, most of which are superconducting copper oxides and related compounds, have been refined routinely with this program.

As well as having many of the 'usual' features of other good programs, such as GSAS and DBWS-9006 described in Chapter 1, RIETAN has some special features of interest. Chief among them are (i) a choice of three different least-squares algorithms (Section 13.1.6), (ii) the possibility to change from one algorithm to another under programmatic control to confirm the convergence to a global minimum (Section 13.1.7), and (iii) incremental refinement in which variable parameters in each cycle are changed appropriately during the course of a refinement series (Section 13.1.7). Databases storing crystallographic information and physical quantities of elements are attached to RIETAN (Section 13.1.2). Both linear and non-linear constraints can be accommodated, the latter along with an exterior penalty function method (Section 13.1.5).

Other convenient features include (i) free input format, (ii) inclusion of comments and labels for parameter groups in input files, and (iii) comprehensive description of linear constraints imposed on refinable parameters (Section 13.1.5). The following conventional features have also been implemented: (i) introduction of 'imaginary' chemical species with average scattering amplitudes of two or more 'real' atoms mixed in arbitrary ratios, (ii) applicability of special symmetry conditions, e.g. in structures with atoms at special positions only (Cooper *et al.* 1977), (iii) usability of non-standard axes setting, (iv) refinement of not the site occupancy multiplier N_j (Chapter 1) but the actual site occupancy, (v) multiphase capability, and (vi) single-pass operation.

In the integrated Rietveld analysis system, FAT–RIETAN, several related programs such as ORTEP-II (Johnson 1976) and ORFFE (Busing *et al.* 1964) are connected to RIETAN via three files storing the results of Rietveld analysis (Fig. 13.1). With further extensions, FAT–RIETAN has grown into a user-friendly interactive system with visually-oriented user interfaces.

Another Rietveld refinement program called PREMOS was also developed in Japan by modifying a REMOS (**RE**finement of **MO**dulated Structures) program for single-crystal X-ray diffraction (Yamamoto *et al.* 1990; Section 13.4). It can not only analyse incommensurate structures and superstructures with powder diffraction data but use sets of data of different kinds in the same refinement (i.e. joint refinement; see Chapter 12). An application of PREMOS in the structure refinement of $Bi_2(Sr_{1-x}Ca_x)_3Cu_2O_{8+z}$ is described in Section 13.5.

13.1.2 *Databases*

The following data, stored in the three formatted sequential files marked #1, #11, and #2 in Fig. 13.1, are invoked by RIETAN as needed during its execution. These contain a few details not present in the databases of similar purpose used in the programs GSAS and DBWS-9006.

1. Laue group numbers, presence/absence of centres of symmetry at the origin, Hermann–Mauguin space group symbols, symmetry conditions, and coordinates of general equivalent positions described for all the cell choices of 230 space groups in *International Tables*, Vol. A (1983).

2. Corresponding descriptions in *International Tables*, Vol. I (1969).

3. Coefficients for analytic approximations of scattering factors and anomalous dispersion corrections listed in *International Tables*, Vol. IV (1974) and Cromer (1976), coherent scattering lengths, incoherent scattering cross-sections, absorption cross-sections (Sears 1986), and atomic weights.

One enters the names of constituent atoms, space group numbers, and setting numbers in conformity with *International Tables*, Vol. I (1969) or Vol. A

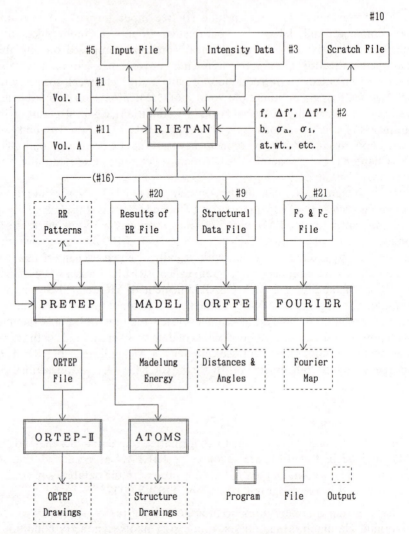

Fig. 13.1 Flow chart of the FAT–RIETAN system. Unit reference numbers are attached to files that are read or created by RIETAN. RR: Rietveld refinement. ATOMS that can be run only on IBM-PC or Macintosh computers is connected to this system via file #9.

(1983). Data stored in the three files can be read by other crystallographic programs.

13.1.3 *Profile shape functions*

13.1.3.1 *Angle-dispersive X-ray and neutron diffraction* The profile shape function, $\phi(2\theta_i - 2\theta_K)$, (see Chapter 1) at the diffraction angle $2\theta_i$ (i: step

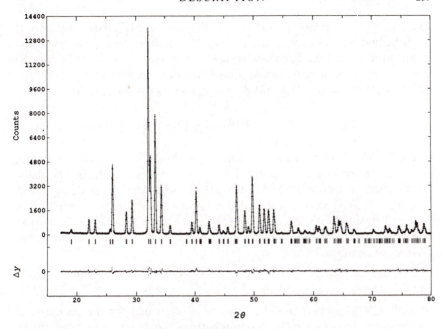

Fig. 13.2 X-ray Rietveld refinement of fluorapatite (CuKα). The solid line is calculated intensities, and points superimposed on it are observed intensities. The differences between the observed and calculated intensities are shown by points appearing at the bottom. Tick marks below the profile indicate the positions of all allowed $K\alpha_1$ and $K\alpha_2$ peaks. The intensity data were kindly supplied by Prof. R. A. Young.

number) for reflection K with the Bragg angle θ_K was represented by Rietveld (1969) as the product of a symmetrical profile shape function, $F(2\theta_i - 2\theta_K)$, and an asymmetrical correction to it, $a(2\theta_i - 2\theta_K)$:

$$\phi(2\theta_i - 2\theta_K) = F(2\theta_i - 2\theta_K)a(2\theta_i - 2\theta_K). \qquad (13.1)$$

As a symmetrical profile shape function in this program, a modified pseudo-Voigt function is adopted in which the Gauss and Lorentz functions may have unequal peak heights and full-width-at-half-maximum intensity (Izumi 1985; Izumi 1989).

This simple profile shape function fits well the Bragg reflection profiles in angle-dispersive X-ray and neutron diffraction patterns of most metals and inorganic compounds of good crystallinity (Fig. 13.2). It is, however, not suited for the fitting of highly asymmetrical profiles. Such profiles can be better approximated by a split function where independent sets of profile parameters are assigned to the left and right sides of peaks (Chapters 7 and 14; Toraya 1986), or by a compound function in which a multi-term Simpson's rule integration is employed (Howard 1982; Von Dreele 1989).

13.1.3.2 *TOF neutron diffraction* A flexible profile shape function is implemented which is optimized for the HRP diffractometer using neutron beams from a solid methane moderator at 20 K (Izumi *et al.* 1987). It is the sum of two exponential profile shape functions proposed by Cole and Windsor (1980), $\phi_1(t_i - t_K)$ and $\phi_2(t_i - t_K)$, in a $(1 - R):R$ ratio:

$$\phi(t_i - t_K) = C\{(1 - R)\phi_1(t_i - t_K) + R\phi_2(t_i - t_K)\}, \tag{13.2}$$

where t_i is the TOF at the ith channel, t_K is the TOF at the peak position, and C is the normalization factor. It may be compared to the profile shape function (eqn (11.7)) described in Chapter 11 which has similar origins but differing details. Observed, calculated, and difference patterns exemplified in Fig. 13.3 demonstrates that eqn (13.2) satisfactorily fits the TOF neutron diffraction data of such a complex solid solution whose peaks are somewhat broadened owing to strain.

13.1.4 *Background corrections*

In RIETAN, as in various other programs, allowance for the background contributions to the overall diffraction pattern may be made in either of two ways:

1. A background function, $B(2\theta_i)$, may be used which is linear in six refinable background parameters $b_0 - b_5$:

$$B(2\theta_i) = \sum_{j=0}^{5} b_j \left(\frac{2\theta_i - \theta_{max} - \theta_{min}}{\theta_{max} - \theta_{min}} \right)^j, \tag{13.3}$$

where θ_{max} and θ_{min} are respectively maximum and minimum θ_i's; therefore, $2\theta_i$ is normalized between -1 and 1 to reduce the correlations among the b_j values.

2. Four pairs of smoothed values of user-selected points in the pattern may be fitted with a power-series polynomial of degree 3, and the value of this polynomial corresponding to the given value of $2\theta_i$ is then calculated.

The use of 1 is preferable to 2 except for samples showing very simple diffraction patterns because peaks overlap to a great extent in a high 2θ region.

    For TOF neutron diffraction data, a background function $B(t_i)$ similar to eqn (13.3) is used, but it is multiplied by the incident intensity and efficiency of counters (Izumi *et al.* 1987). Richardson (Chapter 6) uses a different approach in which amorphous components contributing to the 'background' are represented by a physically-based refinable model.

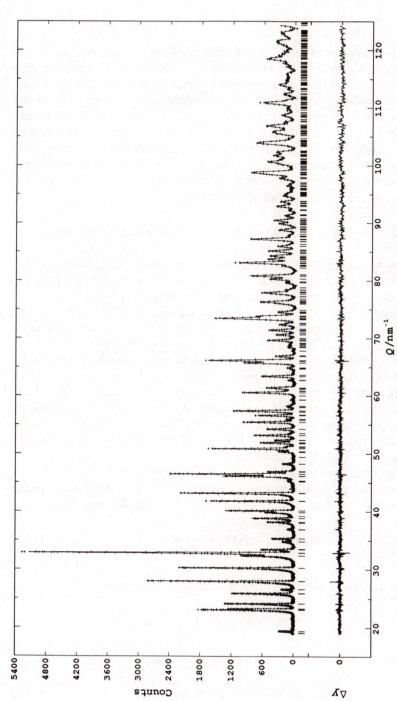

Fig. 13.3 Rietveld refinement patterns for TOF neutron diffraction data of $(Sr_{0.8}Ba_{0.2})(Y_{0.9}Ca_{0.1})(Pb_{0.5}Cu_{0.5})Cu_2O_{7+z}$ (Maeda *et al*, unpublished data). The scattering vector Q ($= 2\pi/d$) is plotted as abscissa and the net intensity as ordinate. Background was fitted as part of the refinement but has been subtracted before plotting.

13.1.5 *Constraints*

13.1.5.1 *Linear constraints* Simple, linear equality constraints can be put in much as Rietveld (1969) did with the use of codewords consisting of a matrix element number and a 'relaxation' factor for each refinable parameter. The 'relaxation' factor is the factor by which the calculated shift for the parameter is to be multiplied before it is applied to construct the new model. In input files for RIETAN, linear constraints can be described in a refined fashion using not codewords but combinations of labels and symbols representing structure parameters. One need not rewrite these constraints even if some parameters are added or deleted, or refinement conditions are partly changed during a series of refinements.

13.1.5.2 *Non-linear constraints* Powder diffraction patterns contain far poorer information than do single-crystal intensity data, mainly because of the overlap of equivalent and neighbouring reflections. Accordingly, a priori information about the crystal structure should be introduced into the program as non-linear inequality and/or equality constraints when reflections overlap very extensively, or when a number of structural parameters have to be refined because many atoms are contained in an asymmetric unit (Chapter 10). Refinement programs developed by Pawley (1980) and Immirzi (1980) eliminate part of the parameters as independent variables by constraining them to be functions of others, or by using derived variables that are more physically meaningful. Since true bond lengths or angles are not exactly equal to expected ones, the introduction of inequality constraints is often preferred to the equality constraints. It also permits the imposition of ranges expected for interatomic distances and angles as well as thermal parameters as boundary conditions.

Constrained non-linear least-squares problems can be solved by an exterior penalty function method (Zangwill 1967). Rietveld analysis under non-linear constraints may be formally stated as (see also eqn 13.7)

$$\text{Minimize:} \quad s(\mathbf{x}) = \sum_i w_i \{y_i - f_i(\mathbf{x})\}^2 \qquad (13.4)$$

subject to p linear and/or non-linear inequality constraints

$$g_j(\mathbf{x}) \geq 0 \qquad j = 1, 2, \ldots, p \qquad (13.5)$$

and q linear and/or non-linear equality constraints

$$h_j(\mathbf{x}) = 0 \qquad j = 1, 2, \ldots, q. \qquad (13.6)$$

In eqns 13.4–13.6, $\mathbf{x}$ is the vector of variable parameters, i is the step number, y_i is the observed intensity, $f_i(\mathbf{x})$ is the calculated intensity, and $w_i \, (= 1/y_i)$ is the weighting based on counting statistics.

The exterior penalty function method transforms a constrained optimization problem into a sequence of unconstrained optimizations for $k = 0, 1, 2, \ldots$ given by

$$\text{Minimize:} \quad S(\mathbf{x}, t^{(k)}) = s(\mathbf{x}) + t^{(k)} \left\{ \sum_{j=1}^{p} H(g_j(\mathbf{x})) g_j^2(\mathbf{x}) + \sum_{j=1}^{q} h_j^2(\mathbf{x}) \right\}, \quad (13.7)$$

where $t^{(k)}$ is a strictly increasing sequence of positive numbers, and H is the Heaviside operator such that $H(x) = 0$ for $x \geq 0$ and $H(x) = 1$ for $x < 0$. The second and third terms in eqn (13.7) prevent the $\mathbf{x}$ vector from getting too far away from the feasible region. A non-feasible sequence of points generated by this algorithm may yield a feasible solution in the limit, i.e. $t^{(k)} = \infty$. Because eqn (13.7) has a sum-of-squares form, conventional algorithms for non-linear least-squares computations, which will be described in Section 13.1.6, can be employed without any modifications.

The algorithm of the exterior penalty function method consists of the following four steps:

1. Set k at 0. Give the initial values of $\mathbf{x}$ and $t^{(0)}$.

2. Refine a set of parameters $\mathbf{x}$ which minimizes $S(\mathbf{x}, t^{(k)})$.

3. If the second and third terms in eqn (13.7) are reduced to nil, stop the calculation since the current values of $\mathbf{x}$ are the solution.

4. Add 1 to k. Increase $t^{(k)}$ and return to step 2.

The non-linear constraints are introduced into the program by means of a separately written function subprogram named CON, whose purpose is to evaluate $H(g_j(\mathbf{x})) g_j(\mathbf{x})$ and $h_j(\mathbf{x})$ when provided with the $\mathbf{x}$ vector. Those partial derivatives of the functionals of constraints with respect to refinable parameters which are used in the set-up of a normal equation (see Section 13.1.6) are approximated by centred differences, not only to save preparation times in formulating analytical derivatives but also to avoid human errors.

13.1.6 *The three non-linear least-squares procedures*

Almost all computer programs for Rietveld refinement employ some form of the Gauss–Newton algorithm to find parameters which minimize $s(\mathbf{x})$ apart from, e.g. XRS-84 (Baerlocher 1984) and MINREF (Elsenhans 1990) which adopt a variable metric method. However, when applied to Rietveld analysis, the Gauss–Newton method suffers disadvantages in that the range of convergence is rather narrow (Howard and Preston 1989) and that refinements often converge to local minima rather than the global minimum (Chapter 3). Since none of the algorithms has proved to be so superior that it can be classified as a panacea for non-linear least-squares solutions, it is advantageous to have more than one method available on call.

In RIETAN, three different techniques are available for non-linear least-squares fitting: the Gauss–Newton method (Nakagawa and Oyanagi 1982), the modified Marquardt method (Fletcher 1971), and the conjugate direction method (Powell 1964). All of them are designed to give stable convergence.

Since the values of initial parameters in an input file can be programmatically replaced with those of final parameters after the refinement, it is very easy to change the method of least-squares at the end of any refinement stage and to continue the calculation from the last point of the previous refinement.

13.1.6.1 *Gauss–Newton method* In this algorithm, changes in n variable parameters at each iterative step, Δx, are calculated by setting up a normal equation:

$$\mathbf{M} \, \Delta \mathbf{x} = \mathbf{N}, \tag{13.8}$$

where $\mathbf{M}$ is the coefficient matrix with n rows and n columns, and both $\Delta \mathbf{x}$ and $\mathbf{N}$ are $n \times 1$ column matrices.

Although $\Delta \mathbf{x}$ is evaluated from $\mathbf{M}^{-1}\mathbf{N}$ in most structure-refinement programs, there is little to recommend such an old-fashioned technique because of long computation time and low precision. In RIETAN, only a lower triangle of the positive-definite symmetric matrix $\mathbf{M}$ is kept in a one-dimensional array to save storage, and the Choleski decomposition of $\mathbf{M}$ and forward- and back-substitutions for the solution of consistent sets of linear equations are carried out (Nash 1979).

A new set of $\mathbf{x}$, $\mathbf{x}'$, is readily obtained by

$$\mathbf{x}' = \mathbf{x} + d \, \Delta \mathbf{x} \tag{13.9}$$

with

$$d = 2^{-m} \qquad m = 0, 1, 2, 3, 4. \tag{13.10}$$

The variable damping factor, d, is initially set at 1 ($m = 0$). If $S(\mathbf{x}', t^{(k)}) > S(\mathbf{x}, t^{(k)})$, d is decreased, and $\mathbf{x}'$ is calculated again with eqn (13.9). The value of d is adjusted appropriately according to the rule adopted in SALS (Nakagawa and Oyanagi 1982).

13.1.6.2 *Modified Marquardt method* This method also calculates **M** and **N** but adds $\lambda \cdot \text{diag}(\mathbf{M})$ (λ: Marquardt parameter, diag: diagonal matrix) to **M** to stabilize the convergence to the minimum:

$$\mathbf{M} + \lambda \cdot \text{diag}(\mathbf{M}) \, \Delta\mathbf{x} = \mathbf{N}. \tag{13.11}$$

Then, $\Delta\mathbf{x}$ tends towards the steepest descents direction as λ becomes larger, while the Gauss–Newton solution is obtained when λ becomes negligible. Even if **M** is not positive definite, it can be made computationally positive definite by choosing λ large enough. The value of λ is automatically adjusted during a series of iterations using a most efficient method developed by Fletcher (1971). Fletcher's algorithm improves the performance of the Marquardt method in certain circumstances, yet requires negligible extra computer time and storage. The modified Marquardt method is very effective for dealing with highly non-linear model functions, $f_i(\mathbf{x})$, or problems in which starting values for parameters differ markedly from the true ones.

13.1.6.3 *Conjugate direction method* The conjugate direction method (Powell 1964) is one of the most efficient algorithms to minimize objective functions without calculating derivatives. The minimum of $S(\mathbf{x}, t^{(k)})$ is located by successive unidimensional searches from an initial point along a set of conjugate directions generated by effective algorithms. In RIETAN, a combination of Davies–Swann–Campey and Powell algorithms (Himmelblau 1972) is adopted as a method of unidimensional minimization. Estimated standard deviations of refinable parameters are obtained by calculating **M** and inverting it after convergence to the solution.

Since the directions for minimization are determined solely from successive evaluations of the objective function, $S(\mathbf{x}, t^{(k)})$, this procedure is much slower than the two least-squares methods with derivatives, but capable of solving ill-conditioned problems in which very high correlations exist between parameters. Because the conjugate direction method is very fast in any 'nearly quadratic' region near a minimum, it is mainly used in the late stages of refinement to test the prospect of a local minimum being the global minimum or to escape from a local minimum by using sufficiently large step sizes of line searches. On the other hand, someone using the Gauss–Newton and Marquardt algorithms can check the convergence to the global minimum simply by using different starting vectors.

13.1.7 *Incremental and combined refinements*

One usually proceeds in steps in Rietveld analysis, first refining only one or two parameters and then gradually letting more and more of the parameters be adjusted in the successive least-squares refinement cycles (Post and

Table 13.1 Convergence behaviour of the Gauss–Newton and modified Marquardt methods in Rietveld analysis of fluorapatite—k: iteration number, l: number of evaluating $s(\mathbf{x})$ in each iteration

	Gauss–Newton method			Marquardt method		
k	$s(\mathbf{x})$	d	l	$s(\mathbf{x})$	λ	l
	1.193×10^6			1.193×10^6		
1^a	3.399×10^5	1.0	1	3.399×10^5	0.0	1
2^b	1.612×10^5	1.0	1	1.612×10^5	0.0	1
3	3.617×10^4	0.5	2	1.588×10^5	8.92×10^{-4}	2
4	7.516×10^3	1.0	1	1.132×10^5	1.78×10^{-3}	1
5	5.704×10^3	0.5	2	1.486×10^4	1.78×10^{-3}	1
6	3.943×10^3	1.0	1	5.977×10^3	8.92×10^{-4}	1
7	3.873×10^3	0.5	2	3.871×10^3	9.74×10^{-4}	2
8	3.725×10^3	1.0	1	3.713×10^3	4.87×10^{-4}	1
9	3.724×10^3	0.5	2	3.346×10^3	4.87×10^{-4}	1
10	3.580×10^3	0.5	2	3.286×10^3	2.44×10^{-4}	1
11	3.333×10^3	1.0	1	3.258×10^3	2.44×10^{-4}	1
12	3.324×10^3	0.25	3	3.257×10^3	2.23×10^{-3}	2
13	3.278×10^3	0.5	1			
14	3.257×10^3	1.0	1			

[a] Refine the scale factor and background parameters.
[b] Refine the FWHM parameters in addition to the above parameters.

Bish 1989). RIETAN requires only a single input to refine parameters incrementally; that is, variable parameters in each cycle can be predesignated by the user or selected appropriately by the program when using the Gauss–Newton and modified Marquardt methods. Repetition of batch jobs is, therefore, unnecessary in most Rietveld refinements. For example, linear parameters (background parameters and a scale factor) are refined in the first cycle, lattice parameters in the second cycle, profile parameters in the third cycle, and subsequently all the parameters simultaneously. Even if initial parameters are far from the true solution, incremental refinements, coupled with the appropriate adjustment of d (Gauss–Newton method) or λ (modified Marquardt method), enable very stable convergence to an optimum solution in most cases (Table 13.1).

Combined refinements are also possible in which the parameters obtained by the incremental refinements described above are further adjusted by the conjugate direction method to ensure that there are no lower minima in the vicinity of the one found by the initial refinement.

13.2 FAT–RIETAN—an integrated system for Rietveld analysis

After the refinement has been finished, three sequential files can be created (Fig. 13.1), i.e. file #20: y_i, $f_i(\mathbf{x})$, and $2\theta_K$ (t_K in TOF neutron diffraction), file #9: final lattice and structural parameters, variance–covariance matrix, symmetry operations, etc., and file #21: hkl, $F_\mathrm{o}\cdot(hkl)$, and $F_\mathrm{c}(hkl)$. Single quotation marks in $F_\mathrm{o}\cdot$ are needed because it is estimated indirectly from summation of the contributions of the peak to net observed intensities (Rietveld 1969).

File #20 is used to draw Rietveld refinement patterns. File #9 is used to calculate interatomic distances, bond angles, root-mean-square components of thermal displacement, etc. with ORFFE (Busing *et al.* 1964), calculate electrostatic site potentials and Madelung energies with MADEL (Kato 1988), and to draw crystal-structure illustrations with ORTEP-II (Johnson 1976) or ATOMS (Dowty 1992). File #21 is used to carry out Fourier and D syntheses with FOURIER (Rotella 1988). Input files for ORTEP-II can be created with macro instructions and file #9 by a preprocessor named PRETEP. Fourier and D syntheses, followed by drawing of contour maps, are very useful for improving structural models in cases where most of the structure has been determined (Cheetham and Taylor 1977). This group of programs including the databases (files #1, #2, and #11) is called the FAT–RIETAN system.

13.2.1 *Interactive FAT–RIETAN system*

The FAT–RIETAN system with visually-oriented user interfaces based on the use of objects such as windows, menus, and icons has been recently developed in collaboration with Rigaku Corporation. One uses a mouse pointer to control these objects and choose one's activities. This system is currently run on a workstation equipped with a bitmapped display and connected, through RS-232C interfaces, to an X-ray powder diffractometer. UNIX system V with multi-task capability and a window system is used as an operating system to run multiple application programs and allow the screen to display them at the same time. Various members of FAT–RIETAN such as RIETAN, ORFFE, and ORTEP-II can be executed simultaneously during collection of X-ray intensity data. One can resize, hide, or expose a window with the mouse. Since about 8000 lines of program code were written in the C language to make the FAT–RIETAN system interactive and user-friendly, it will be readily usable by non-experts who are not familiar with commands, text editing, and file manipulations in UNIX.

This interactive and integrated system can also be used as a powerful tool for computer-aided instruction (CAI) of crystallography, where raw data taken on a conventional X-ray powder diffractometer are analysed with RIETAN. FAT–RIETAN includes a fairly large part of the crystallo-graphic calculations used in single-crystal X-ray analysis except for phase determination.

13.3 Further RIETAN developments

The program written for the fixed-wavelength case includes the capability for Rietveld analysis of synchrotron X-ray data, e.g. those obtained from a dedicated powder diffractometer set up at the Photon Factory at KEK (Ozawa *et al.* 1989). Honda *et al.* (1990) collected synchrotron X-ray data by using a large-radius (28.65 cm) Weisenberg camera equipped with a Fuji Imaging Plate and succeeded in the *ab initio* structure determination of 5-aminovaleric acid. RIETAN was used to refine the structure parameters of this organic compound in the final stage of its structure analysis. The agreement between observed and calculated data was very good in spite of the facts that the peak shape was simply assumed to be symmetric Gaussian and that no preferred-orientation correction was used.

As in the GSAS program described in Chapters 1 and 12, RIETAN is applicable to the refinement of crystal structures from powder data obtained from four different methods (Albinati and Willis 1982). These are fixed-wavelength methods with the radiation from laboratory sources of characteristic X-rays, from synchrotron X-ray sources, and from steady-state nuclear reactor neutron sources, plus fixed-angle (varying wavelength) methods with neutrons from pulsed sources. The DBWS-9006 program mentioned in Chapter 1 is applicable only to the three fixed-wavelength methods.

13.4 Rietveld analysis of incommensurate structures

The Rietveld method may be applied to sophisticated analysis which has been hitherto regarded as almost impossible. Recently, the PREMOS program has been developed which can refine incommensurate structures as well as superstructures by the Rietveld method (Yamamoto *et al.* 1990). It also makes possible joint refinement with X-ray and neutron diffraction data under non-linear constraints. The algorithms adopted in REMOS for the analysis of single-crystal intensity data (Yamamoto 1982) have been com-bined with the Rietveld method. Elsenhans (1990) independently developed a MINREF program for Rietveld analysis of incommensurate nuclear and magnetic structures with neutron powder data, but neither fractional coordinates nor occupation factors are refinable in the present version.

Peak positions and structure factors for (one-dimensional) incommen-surate structures are calculated in more complicated ways than for commen-surate ones. Four integers, *hklm*, are needed to index main and satellite peaks systematically. The reciprocal-lattice vector, $\mathbf{q}$, can be written in vector notation

$$\mathbf{q} = h\mathbf{a}^* + k\mathbf{b}^* + l\mathbf{c}^* + m\mathbf{k} \tag{13.12}$$

with

$$\mathbf{k} = k_1\mathbf{a}^* + k_2\mathbf{b}^* + k_3\mathbf{c}^*, \tag{13.13}$$

where $\mathbf{k}$ is the wave vector of the modulation wave, and $\mathbf{a}^*$, $\mathbf{b}^*$, and $\mathbf{c}^*$ are reciprocal unit-cell vectors for the subcell. Then, the interplanar spacing, d, can be obtained by

$$d = |\mathbf{q}|^{-1}. \tag{13.14}$$

For example, d is expressed simply as

$$d = \{(h + mk_1)^2 a^{*2} + (k + mk_2)^2 b^{*2} + (l + mk_3)^2 c^{*2}\}^{-1/2} \tag{13.15}$$

in cubic, tetragonal, and orthorhombic forms.

Additional parameters are also necessary to calculate structure factors for incommensurate structures. For example, the atomic position, $\mathbf{r}$, is calculated by adding cosine and sine waves to the average position, $\bar{\mathbf{r}}$:

$$\mathbf{r} = \bar{\mathbf{r}} + \mathbf{u}_c \cos(2\pi t) + \mathbf{u}_s \sin(2\pi t), \tag{13.16}$$

where $\mathbf{u}_c$ and $\mathbf{u}_s$ are respectively the amplitudes of the cosine and sine waves, and $t\ (=\mathbf{k}\cdot\bar{\mathbf{r}})$ is the phase of the wave. The occupation factor and isotropic thermal parameter can be expressed by similar equations.

13.5   Rietveld refinement of the modulated structure for $Bi_2(Sr_{1-x}Ca_x)_3Cu_2O_{8+z}$

Several models have been proposed for the incommensurate modulated structure of a high-T_c superconductor $Bi_2(Sr_{1-x}Ca_x)_3Cu_2O_{8+z}$ ($T_c \approx 80$ K), but none of them have succeeded in explaining all the structural data obtained by high-resolution transmission electron microscopy (HRTEM), X-ray and neutron diffraction. Since it contains Bi as a principal component, reliable structural parameters of oxygen can hardly be obtained without the use of neutron diffraction data. In fact, even single-crystal X-ray analysis (Gao et al. 1988) failed in refining the atomic displacement amplitudes of oxygen atoms. X-ray powder data are also needed to precisely evaluate the distribution of the metal atoms among some metal sites.

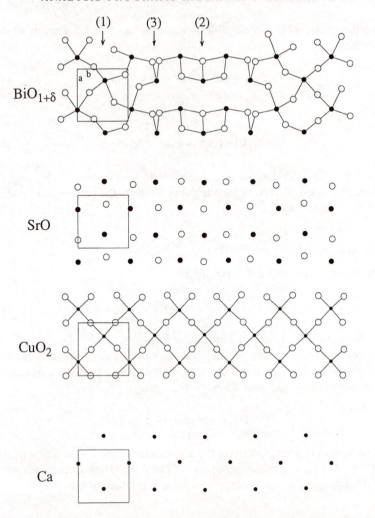

Fig. 13.4 Four kinds of sheets projected on the *ab* plane in the modulated structure of $Bi_2(Sr_{1-x}Ca_x)_3Cu_2O_{8+z}$. Metal atoms are shaded. Bi—O bonds shorter than 0.27 nm are connected. Oxygen atoms in the transient region are split into two fragments.

The incommensurate structure of $Bi_2(Sr_{1-x}Ca_x)_3Cu_2O_{8+z}$ was refined by combining X-ray and TOF neutron powder diffraction data (Yamamoto *et al.* 1990). In this superconductor, Bragg reflections have satellite pairs parallel to $\mathbf{b}^*$, and the incommensurate periodicity is estimated as $4.8b$ by electron diffraction (Onoda *et al.* 1988). Therefore, the wave vector, $\mathbf{k}$, is approximately equal to $\mathbf{b}^*/4.8$. Anisotropic broadening of peaks due to syntactic intergrowths made it hard to fit calculated patterns to observed ones satisfactorily. However, structural details such as oxygen

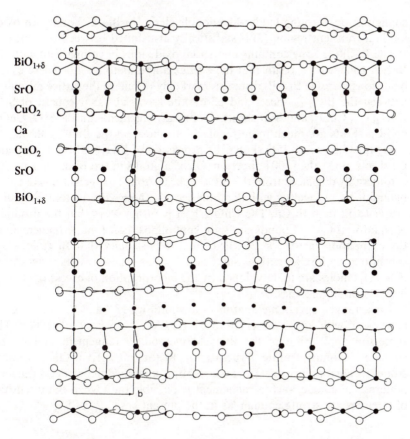

Fig. 13.5 Projection view on the *bc* plane in the modulated structure of $Bi_2(Sr_{1-x}Ca_x)_3$-Cu_2O_{8+z}. Metal atoms are shaded. Cu—O bonds shorter than 0.25 nm and Bi—O bonds shorter than 0.27 nm are connected.

arrangements on $BiO_{1+\delta}$ sheets have been successfully clarified for the first time.

HRTEM (high resolution transmission electron microscopy) revealed that Bi-dilute and Bi-dense zones alternate along the [010] direction (Matsui *et al.* 1988). As Fig. 13.4 (see p. 250) shows, each Bi cation in the Bi-dilute zone (1) is coordinated to four oxide ions on the $BiO_{1+\delta}$ sheet and one oxide ion on an adjacent SrO sheet. This Bi atom has a chemical environment similar to that of a B site cation in a perovskite-type compound, except that no oxygen atoms are sandwiched between two $BiO_{1+\delta}$ sheets. On the other hand, each Bi cation in the Bi-rich zone (2) is coordinated to four oxide ions, i.e. to three oxide ions on the $BiO_{1+\delta}$ sheet and one oxide ion on the SrO sheet. The Bi and O atoms in this zone have a distorted NaCl-type configuration, forming ladder-like double chains of Bi—O bonds. An oxide ion in a transient region (3) is statistically distributed between two sites.

The Bi-dilute zone contains interstitial oxide ions in comparison with the Bi-rich zone, with a result that the z value amounts to *ca.* 1.0. These excess oxide ions increase Bi—Bi distances markedly to relieve the lattice mismatch between the $BiO_{1+\delta}$ sheets and perovskite-type slabs that consist of SrO, CuO_2, Ca, CuO_2, and SrO sheets. Atomic arrangements on the SrO, CuO_2, and Ca sheets are much more regular than those on the $BiO_{1+\delta}$ sheet.

Figure 13.5 (see p. 251) shows the projection of $BiO_{1+\delta}$, SrO, CuO_2, and Ca sheets along the [100] direction. Oxygen atoms in the Bi-dilute zone are considerably displaced from the $BiO_{1+\delta}$ sheet. Apical oxygen atoms of CuO_5 pyramids move together with Bi atoms because these oxygen atoms are much nearer to Bi than to Cu. The CuO_2 sheet is rather wavy, but the distances of equatorial Cu—O bonds on it are kept almost constant. A refinement of the occupation factor for Bi ruled out the possibility that Bi cations are substituted by foreign cations such as Sr^{2+} ions. These results suggest that $(Cu—O)^+$ holes are mainly doped by the incorporation of excess oxide ions into the Bi-dilute zones.

The arrangements of metal atoms shown in Figs 13.4 and 13.5 are in good agreement with those observed by HRTEM (Matsui *et al.* 1988). The successful application of the Rietveld method to refinement of the very complex incommensurate structure of $Bi_2(Sr_{1-x}Ca_x)_3Cu_2O_{8+z}$ demonstrates its great ability to extract information contained in powder diffraction patterns. Of course, such a refinement is possible only when the intensities of satellites are strong enough, as in the present case.

References

Albinati, A. and Willis, B. T. M. (1982). *J. Appl. Crystallogr.*, **15**, 361–74.

Baerlocher, Ch. (1984). *Acta Crystallogr.*, **A40**, C-368.

Busing, W. R., Martin, K. O., and Levy, H. A. (1964). Report ORNL-TM-306, Oak Ridge National Laboratory, Tennessee.

Cheetham, A. K. and Taylor, J. C. (1977). *J. Solid State Chem.*, **21**, 253–75.

Cole, I. and Windsor, C. G. (1980). *Nucl. Instrum. Methods*, **171**, 107–13.

Cooper, M. J., Rouse, K. D., and Sayers, R. (1977). AERE-R8695, AERE Harwell.

Cromer, D. T. (1976). *Acta Crystallogr.*, **A32**, 339.

Dowty, E. (1992). *ATOMS: a computer program for displaying atomic structures.* Shape Software, Kingsport.

Elsenhans, O. (1990). *J. Appl. Crystallogr.*, **23**, 73–6.

Fletcher, R. (1971). AERE-R6799, AERE Harwell.

Gao, Y., Lee, P., Coppens, P., Subramanian, M. A., and Sleight, A. W. (1988). *Science*, **241**, 954–6.

Himmelblau, D. M. (1972). *Applied nonlinear programming*, pp. 42–9. McGraw-Hill, New York.

Honda, K., Goto, M., and Kurahashi, M. (1990). *Chem. Lett.*, **1990**, 13–16.

Howard, C. J. (1982). *J. Appl. Crystallogr.*, **15**, 615–20.
Howard, S. A. and Preston, K. D. (1989). In *Modern powder diffraction*, Reviews in mineralogy, Vol. 20 (ed. D. L. Bish and J. E. Post), pp. 217–75. Mineralogical Society of America, Washington, DC.
Immirzi, A. (1980). *Acta Crystallogr.*, **B36**, 2378–85.
International tables for crystallography (1983). Vol. A. D. Reidel, Dordrecht.
International tables for X-ray crystallography (1969). Vol. I. Kynoch Press, Birmingham.
International tables for X-ray crystallography (1974). Vol. IV. Kynoch Press, Birmingham.
Izumi, F. (1985). *Nippon Kessho Gakkai Shi* (*J. Crystallogr. Soc. Jpn.*) (in Japanese), **27**, 23–31.
Izumi, F. (1989). *Rigaku J.*, **6**, No. 1, 10–19.
Izumi, F., Asano, H., Murata, H., and Watanabe, N. (1987). *J. Appl. Crystallogr.*, **20**, 411–18.
Johnson, C. K. (1976). Report ORNL-5138, Oak Ridge National Laboratory, Tennessee.
Kato, K. (1988). Unpublished program.
Maeda, T. *et al.* Unpublished data.
Matsui, Y., Maeda, H., Tanaka, Y., and Horiuchi, S. (1988). *Jpn. J. Appl. Phys.*, **27**, L372–5.
Nakagawa, T. and Oyanagi, Y. (1982). *Analysis of experimental data with least-squares methods* (in Japanese), pp. 98–101. Tokyo Daigaku Shuppankai, Tokyo.
Nash, J. C. (1979). *Compact numerical methods for computers: linear algebra and function minimisation*, pp. 72–5. Adam Hilger, Bristol.
Onoda, M., Yamamoto, A., Takayama-Muromachi, E., and Takekawa, S. (1988). *Jpn. J. Appl. Phys.*, **27**, L833–6.
Ozawa, H., Uno, R., Yamanaka, T., Morikawa, H., Ando, M., Ohsumi, K., *et al.* (1989). *Rev. Sci. Instrum.*, **60**, 2382–5.
Pawley, G. S. (1980). *J. Appl. Crystallogr.*, **13**, 630–3.
Post, J. E. and Bish, D. L. (1989). In *Modern powder diffraction*, Reviews in mineralogy, Vol. 20 (ed. D. L. Bish and J. E. Post), pp. 277–308. Mineralogical Society of America, Washington, DC.
Powell, M. J. D. (1964). *Computer J.*, **7**, 155–62.
Rietveld, H. M. (1969). *J. Appl. Crystallogr.*, **2**, 65–71.
Rotella, F. J. (1988). *Users manual for Rietveld analysis of time-of-flight neutron powder diffraction data at IPNS.* Argonne National Laboratory, Illinois.
Sears, V. F. (1986). In *Neutron scattering (Part A)*, Methods of experimental physics, Vol. 23 (ed. K. Sköld and D. L. Price), pp. 521–47. Academic Press, New York.
Toraya, H. (1986). *J. Appl. Crystallogr.*, **19**, 440–7.
Von Dreele, R. B. (1989). In *Modern powder diffraction*, Reviews in mineralogy, Vol. 20 (ed. D. L. Bish and J. E. Post), pp. 333–69. Mineralogical Society of America, Washington, DC.
Watanabe, N., Asano, H., Isawa, H., Satoh, S., Murata, H., Karahashi, K. *et al.* (1987). *Jpn. J. Appl. Phys.*, **26**, 1164–9.
Yamamoto, A. (1982). *Acta Crystallogr.*, **A38**, 87–92.
Yamamoto, A., Onoda, M., Takayama-Muromachi, E., Izumi, F., Ishigaki, T., and Asano, H. (1990). *Phys. Rev.*, **B42**, 4228–39.
Zangwill, W. I. (1967). *Manage. Sci.*, **13**, 344–58.

14

Position-constrained and unconstrained powder-pattern-decomposition methods

Hideo Toraya

14.1 Introduction
14.2 Theory
14.3 Applications of the individual profile-fitting method
14.4 Applications of the Pawley method
14.5 Conclusion
14.6 Acknowledgement
References

14.1 Introduction

The individual profile fitting, the Pawley, and the Rietveld methods are compared in Table 14.1. The first two methods, described in this chapter, are classified as pattern decomposition methods. With them, one executes the decomposition of the powder pattern into individual Bragg components without reference to a structural model (Young *et al.* 1982). The difference between these two models is that the peak positions are unconstrained, varied independently as adjustable parameters in the least-squares in the individual profile fitting method while they are constrained by adjustable unit-cell parameters in the Pawley method. Furthermore, in individual profile fittings, pattern decomposition is carried out in a small 2θ-range, changing the range of analysis successively, whereas the whole powder pattern is decomposed in one step by the Pawley method. These two methods do not require crystal structural data (atom site occupancy, positional, and thermal parameters) to start the least-squares fitting. The knowledge required initially is (i) the number of peaks in the analysed range in the individual profile fitting, and (ii) the approximate unit-cell parameters in the Pawley method.

Historically, the problem of overlapping reflections in powder diffraction has been discussed in a number of publications. Some were concerned with the 'search-match' or the indexing of powder diagrams, and some with structure analysis with powder diffraction data. Mortier and Constenoble (1973) proposed a method of decomposing the overlapping reflections in a limited 2θ-range by using an experimentally determined Fourier series. Just prior to their work, pattern decomposition method using least-squares fitting of an analytical-type profile function was proposed by Taupin (1973), which

Table 14.1 Comparison of methods

	Individual profile-fitting method	Pawley method	Rietveld method
Aim of analysis	Pattern decomposition	Pattern decomposition and refinement of unit cell parameters	Structure refinement
Range of analysis	Partial patterns	Whole pattern	Whole pattern
Profile model { Profile area	Independent parameters	Independent parameters	Function of structural parameters
Peak position	Independent parameters	Function of unit-cell parameters	Function of unit-cell parameters
Profile shape	Independent of angle in small 2θ range	Angle-dependent	Angle-dependent
A priori knowledge required to start the refinement	Null	Approximate unit-cell parameters	Initial cell and structural parameters

was further developed by Parrish's group (Huang and Parrish 1975; Parrish *et al.* 1976). The profile function used in those studies was the convolution of a single Lorentzian function, to represent the specimen effect, with a pre-determined instrumental profile which was represented by the sum of seven Lorentzian functions. The most often used symmetric profile functions are shown in Table 1.3. Sonneveld and Visser (1975) used a single modified Lorentzian function for representing the observed diffraction profile for pattern decomposition with Guinier camera data. In the mid-1970s, the Rietveld method (Rietveld 1969) was extended to the analysis of X-ray powder diffraction data (Mackie and Young 1975; Malmros and Thomas 1977; Young *et al.* 1977; Khattak and Cox 1977). Since then the use of flexible profile functions such as the Pearson VII (Hall *et al.* 1977) have spread, and nowadays, single profile functions are widely used for approximating the observed diffraction profile, instead of a convolution of two or more functions, in individual profile fitting. (But see Chapter 7.) The use of an explicit convolution of functions is, however, again important in extracting microstructural information such as crystallite size and microstrain from overlapping reflections. This problem is discussed later (Sections 14.3.2 and 14.4.3 and also in Chapter 8 relative to pattern decomposition). Dozens of computer programs have been written for fitting individual profiles, and some can be run on small personal computers. Such computer programs are installed by the manufacturers as standard data analysis routines in some commercially available powder diffractometers.

The Pawley method (Pawley 1981) was first proposed as a procedure for refining unit-cell parameters and providing a list of indexed integrated intensities. The computer program is similar, in its style, to that for the Rietveld method except for the refinement of structural parameters. In fact, as described in Pawley's paper, the computer program ALLHKL for the Pawley method, first written for analysing fixed-wavelength neutron powder diffraction data, was modified from that for the Rietveld method by removing the routine for calculating the intensities from an assumed structure (Pawley 1981). The method was then extended to the analysis of X-ray powder diffractometer data (the computer program WPPF), where the modelling of angle-dependent profile shape and the correction for peak-shift arising largely from the peak asymmetry are more important than they are in neutron powder diffraction (Toraya 1986). The other computer programs for applying the Pawley method, in so far as the author knows, are PROFIT (Scott 1987) and FULFIT (Jansen *et al.* 1988). The Pawley method is powerful in providing the structure factors for the Patterson or direct methods in *ab initio* structure determination with powder diffraction data and thus complements the Rietveld method as a combined technique for the structure solution and refinement. Other important applications of the Pawley method are found in the field of materials science, where the measurement of unit-cell parameters and/or microstructural analysis are routine tasks; the method is best suited to these analyses.

In this chapter, the theories of the individual-profile-fitting and Pawley methods in the angle-dispersive case, and their applications, are described.

14.2 Theory

14.2.1 *Fitting function*

As is described in Chapter 1, the experimental diffraction profile $h(x)$ is the convolution of the intrinsic diffraction profile $f(x)$ with the instrumental function $g(x)$, and is expressed by (eqn (1.7))

$$h(x) = \int f(x - x')g(x')\,dx', \qquad (14.1)$$

where x and x' are 2θ angles. The function $f(x)$ represents the profile broadening arising from the specimen, and the $g(x)$ includes both spectral distribution and geometrical instrumental factors. The distribution of intensity for the total powder pattern $y(x)$, can be expressed by (as in eqn (1.2))

$$y(x) = b(x) + \sum_{K} h(x)_{K}, \qquad (14.2)$$

where $b(x)$ is the background intensity (see Chapter 1) and the summation $\sum_K$ is taken over all K reflections ($K = hkl$) which contribute at x. The function $h(x)_K$ by (14.1) can be represented

$$h(x)_K \cong I_K \phi(x - x_K)_K, \tag{14.3}$$

where $\phi(x)_K$ is the normalized profile function used to model the diffraction profile of $h(x)_K$, x_K is the peak maximum position, and I_K represents the integrated intensity. From (14.2) and (14.3), the $y(x)$ at the ith point is

$$y(x_i) = b(x_i) + \sum_K I_K \phi(x_i - x_K)_K. \tag{14.4}$$

Chapter 1 supplements the description of the procedure of least-squares fitting and the definition of R factors (Table 1.3).

14.2.2 *Integrated intensity parameter*

Equation (14.4) is commonly used as a fitting function in the three methods classified in Table 14.1. It is identical with eqn (1.2) in Chapter 1 if we replace the integrated intensity parameter I_K with the squared absolute magnitude of structure factor multiplied by the correction factors for Lorentz-polarization and preferred orientation. In the individual-profile-fitting and Pawley methods, the I_K's in eqn (14.4) are, in principle, varied independently in the least-squares fitting. However, (i) ill-conditioning of the least-squares matrix occurs for very closely-spaced overlapping reflections, and (ii) the matrix becomes singular when reflections are at the same Bragg reflection position (intrinsic overlapping). In the latter case, those reflections are regarded as one reflection, and one variable parameter is assigned in least-squares fitting. In the former problem, most computer programs currently used for individual profile fitting are not equipped to handle this problem, and either (i) the intensities are fixed to avoid parameter divergence during the least-squares or (ii) fitting is abandoned for that small 2θ range. However, in the Pawley method when fitting the whole powder pattern, ill-conditioning in part of the matrix results in failure as a whole. Pawley (1981) introduced the concept of grouping of reflections and used a slack constraint on the variation of the integrated intensity parameters belonging to that group. Other programs for this type of refinement also use the grouping strategy, although hard constraints (fixed intensity ratio, for example, of 1:1 for group members) are also used (Toraya 1986; Scott 1987).

Another problem in refining all integrated intensity parameters in the whole pattern is that the size of the matrix becomes very large. If there are, for example, 200 reflections in the analysed 2θ range, we must invert the matrix with $(200 + P)$ elements, where P is the number of profile and unit-cell parameters. An interesting feature of Pawley's refinement is that there is no parameter interaction for I_K's between two reflections well apart from each other (non-overlapping), and thus most elements of the matrix consisting of $\sum_i w_i[\partial y(x_i)/\partial I_K][\partial y(x_i)/\partial I_{K'}]$ are zero (for example, it amounts to about 70 per cent of the total matrix elements when the number of independent reflections = 200). Computer memories and computation time have been saved by replacing a large triangular matrix with two narrow band matrices for non-zero elements (Toraya 1986).

14.2.3   Peak positions

An important feature of the Pawley method is that the peak positions of individual reflections are constrained by the unit-cell parameters. The position x_K of Bragg reflection hkl is calculated from the reciprocal lattice parameters a^*, b^*, c^*, α^*, β^*, and γ^* by

$$x_K = 2\sin^{-1}(\lambda d^*_{hkl}/2),$$

$$d^*_{hkl} = [h^2 a^{*2} + k^2 b^{*2} + l^2 c^{*2} + 2kl b^* c^* \cos\alpha^* + 2lh c^* a^* \cos\beta^*$$
$$+ 2hk a^* b^* \cos\gamma^*]^{1/2}, \tag{14.5}$$

where λ is the wavelength. The reciprocal or direct lattice parameters are refined during least-squares fitting. This constraint requires that approximate unit-cell parameters be used to start the least-squares. In most cases, however, we can easily find out the rough parameter values from crystallographic databases such as the Powder Diffraction File of the JCPDS-ICDD or Structure Reports, and in the remaining cases, we will have to rely on the computer programs for indexing powder patterns. Position-constrained pattern decomposition by the Pawley method is much more robust in the least-squares procedure than is position-unconstrained individual profile fitting.

14.2.4   Profile function

A profile function with a poor approximation to the observed diffraction profile causes the derived parameter values to become less accurate and sometimes induces oscillation of the parameters and parameter divergence in least-squares cycles. This problem is more severe in the less constrained individual-profile-fitting method. The Pearson VII (Hall *et al.* 1977), the split Pearson VII (Chapter 7), the Voigt (Appendix 8A to Chapter 8; Voigt 1912; Langford 1978), and the pseudo-Voigt (Wertheim *et al.* 1974) functions have been demonstrated to give the best fit to the observed X-ray diffraction profile among the profile functions available for testing (Young and

Wiles 1982; Langford 1987), and they have been widely used in both individual profile fitting and Pawley methods as well as in the Rietveld method.

Profile modelling is difficult for profile asymmetry, which tends to be large in the low angle region, particularly, in the case of the commonly-used $\theta-2\theta$ X-ray diffractometer (see Chapters 7 and 9). Several methods devised for modelling the profile asymmetry are: (i) to multiply a symmetric function by an asymmetry factor, which is greater and less than unity on the low- and high-angle sides, respectively (Rietveld 1969; Will et al. 1987), (ii) to dispose asymmetrically two or three symmetric functions, for example, of Lorentzian type proposed by Parrish et al. (1976) for the X-ray case, or of Gaussian type proposed in the case of neutron diffraction (Howard 1982), (iii) to use a split-type function, consisting of two functions, each of which is solely defined on the low-angle side or on the high-angle side of the peak maximum (Windsor and Sinclair 1976 and see Chapter 7), and (iv) to convolute a symmetric function for $f(x)$ with an asymmetric profile for $g(x)$ (Taupin 1973; Howard & Snyder 1984; Toraya 1988). These modellings are based on the asymmetrization of the symmetric (analytical-type) profile function. Another approach is to use an empirical function such as a tabulated observed diffraction profile (Hepp and Baerlocher 1988). A new array-type profile function can assume any complex shape and has been shown to be amenable to least-squares fitting to very asymmetric profiles (Toraya 1990; Toraya and Parrish 1990).

In whole-powder-pattern fitting, the angle-dependency of the profile shape must be incorporated into the profile model. The variation of FWHM with 2θ is expressed, as in many computer programs for the Rietveld method in the angle-dispersive case, by the Caglioti et al. formula (1958) initially derived for neutron powder diffraction (eqn (1.6)).

The asymmetry factor proposed by Rietveld (1969) is dependent on $(\tan \theta)^{-1}$. This empirical formulation is widely used in the angle-dispersive neutron powder diffraction case, as in ALLHKL. In the case of angle-dispersive X-ray data, the diffraction profile has a more Lorentzian shape and much greater asymmetry, and it is reported that the incorporation of a fourth order term into the asymmetry factor has improved the fitting (Langford et al. 1986). In the case of a split-type function, the peak asymmetry is defined as $A = H_l/H_h$, where H_l and H_h are the FWHM on the low- and high-angle sides, respectively (functional forms for normalized split Pearson VII and pseudo-Voigt functions are given in Toraya 1990). The parameter A is the adjustable parameter in the case of individual profile fitting, while it is angle-dependent in the whole-powder-pattern fitting case, having a form such as

$$A(2\theta) = a_1 + a_2(2^{\frac{1}{2}} - 1/\sin \theta) + a_3(2 - 1/\sin^2 \theta), \qquad (14.6)$$

where a_1, a_2, and a_3 are adjustable parameters (Toraya 1986). In the case of a convolution of functions, the angle-dependency of the $g(x)$ profile is provided by choosing the observed diffraction profiles at various 2θ angles. The parameters defining the profile shape such as η in the pseudo-Voigt and the exponent, m, in the Pearson VII function (Table 1.2) are also known to vary with 2θ angle, and 2θ-dependent functions have been devised (Brown and Edmonds 1980; Hill and Howard 1985; Toraya 1986; and see Chapter 7).

14.3   Applications of the individual profile-fitting method

14.3.1   *Individual profile fitting with constraints on profile shape*

In individual profile fitting, the integrated intensity, the peak position, and the profile shape parameters are assigned to each reflection in a selected 2θ-range of analysis. In principle, the reflections are independent entities, and all parameters are refined independently along with those for the background. In practice, however, the parameters of profile shape and width for neighbouring reflections are usually constrained in order to avoid ill-conditioning of the least-squares matrix. An interesting example of how easily things can go wrong in this kind of profile fitting without constraints, *especially* if asymmetric profiles are used, is given in Chapter 7 by Figs 7.5 and 7.6.

The following is an example of the use of constraints. The sample was a mixture of α-SiO_2 and Si in 2:1 weight ratio. The data for three overlapping reflections (105/015 and 401/041 for α-SiO_2 and 333/511 for Si), collected with a conventional diffractometer and Cu K_α radiation, were analysed with the computer program PROFIT using split-type Pearson VII functions (Toraya 1986). Table 14.2 gives a comparison of two refinement results, A and B, obtained under the assumption that in case A the three profiles had the same shape and width and that in case B the profiles had the same shapes but different widths for α-SiO_2 and Si. The fitting result of refinement B is shown in Fig. 14.1. In refinement B there was a small (0.5 per cent) decrease in R_p and R_{wp} factors compared to refinement A, and the FWHM were $0.136(3)°$ for α-SiO_2 and $0.171(2)°$ for Si, while the common profile breadth was refined to $0.157(2)°$ in refinement A. The integrated intensities in B also significantly changed from those in A. Is the difference in profile widths significant? The FWHM were $0.132°$ for α-SiO_2 and $0.171°$ for Si, which were obtained separately from individual profile fittings for single phases of α-SiO_2 and Si (Table 14.2), and the refinement B did resolve the difference in this case.

Pattern decomposition becomes difficult for reflections that are weak, broadened, closely overlapping, and have high statistical noise. More constraints are then required in the least-squares procedure. The judgement of whether a constraint is necessary for this and that profile, or not, is heavily

Table 14.2 Refined least-squares parameters, R_p and R_{wp} factors in refinements A and B

		Refinement A			Refinement B	
$b_0{}^a$		$-30(68)$			$-86(56)$	
b_1		$0.6(7)$			$1.1(6)$	
hkl	105/015	333/511	401/041	105/015	333/511	401/041
I_K	44(1)	113(2)	43(1)	40(1)	130(3)	35(1)
t_K (°)	94.619(2)	94.923(1)	95.088(2)	94.613(1)	94.923(1)	95.091(1)
H_K (°)		0.157(2)		0.136(3)	0.171(2)	0.136
A^b		1.33(6)			1.20(5)	
m_l		1.47(9)			1.21(5)	
m_h		1.09(6)			1.07(4)	
H (°)^c			0.132		0.171	0.132
R_p (%)		1.8			1.3	
R_{wp} (%)		2.4			1.8	

^{a} b_0 and b_1 are adjustable parameters in the background function $b(2\theta) = b_0 + b_1 2\theta$.
^{b} A, m_l, and m_h are the asymmetry parameter and the exponents in the Pearson VII function on the low- and high-angle sides, respectively.
^{c} H's are obtained separately from profile fittings for single phases of α-SiO$_2$ and Si.

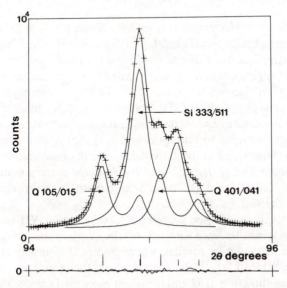

Fig. 14.1 Fitting result for three overlapping reflections from α-SiO$_2$ and Si. Observed and calculated profile intensities are represented by symbol $+$ and solid line, respectively. Differences between the two intensities are plotted at the bottom of the diagram on the same scale as above. Short vertical bars represent Bragg reflections positions for Kα_1 and Kα_2.

dependent on the program-user's experiences and on 'trial and error' at present. The important thing is to obtain a physically meaningful result rather than the lowest R factors. Future computer programs will probably be able to make such 'judgements'.

14.3.2   Use of a convolution of functions in individual profile fitting

The procedure of decomposition is used to resolve the pattern of superimposed profiles into the individual component profiles $h(x)_K$. On the other hand, a deconvolution procedure is required to unfold the integral equation (14.1), that is, to derive $f(x)$ (or $g(x)$) from $h(x)$ and $g(x)$ (or $f(x)$). The problem of deconvoluting X-ray diffraction profiles has a long history, and its detailed description is beyond the scope of this chapter. Stoke's (1948) method involving Fourier series and direct fitting methods (Louër et al. 1969; Moraweck et al. 1977) are known to give precise results. An excellent example of determining the shape, size, and size distribution of ZnO crystallites by the Fourier (Warren 1969) and variance (Wilson 1963) methods was given by Louër et al. (1983). However, these techniques can be applied only to isolated peaks, and the range of materials to be so analysed is, therefore, very limited. The individual-profile-fitting technique makes it possible to determine the individual diffraction profiles in an observed set of overlapping reflections.

A first approach to this problem is to use a Voigt function for measuring line broadening as the integral breadth β or the FWHM. The Voigt function (V), being a convolution of Lorentzian (L) and Gaussian (G) functions, gives a much better approximation to X-ray diffraction profiles than do single Lorentzian or Gaussian functions. Langford (1978) gives an explicit way for deriving the integral breadth of the Voigt function, β_V, from the integral breadths of its constituent Lorentzian and Gaussian components, β_L and β_G. Fitting of Voigt functions to both $g(x)$ and $h(x)$ thus allows determination of the breadth parameters of the Lorentzian and the Gaussian components of the Voigt function, H_{Lg}, H_{Lh}, H_{Gg}, and H_{Gh} (and the corresponding integral breadths β_{Lg}, β_{Lh}, β_{Gg}, and β_{Gh}). The integral breadth of the specimen profile $f(x)$, β_{Vf}, is derived by $\beta_{Lf} = \beta_{Lh} - \beta_{Lg}$ and $\beta_{Gf}^2 = \beta_{Gh}^2 - \beta_{Gg}^2$, and the formula mentioned above (see Chapter 8). The Pearson VII and pseudo-Voigt functions can be used instead of the Voigt function as a good approximation for Voigtian profiles, where empirical formulae are used for deducing the FWHM of the Voigt function from the H and η parameters of the pseudo-Voigt function or H and exponent m, of the Pearson VII function (De Keijser et al. 1982).

The second approach is based on the following function, which is obtained from eqns (14.1) and (14.2).

$$y(x_i) = b(x_i) + \sum_K \int f(x_i - x)_K g(x)_K \, dx. \tag{14.7}$$

An analytical-type function is assumed for $f(x)$, and the function (analytical- or empirical-type), which is pre-determined from measurement on a standard reference material, is used for $g(x)$. The convolution in eqn (14.7) is carried out by direct integration, when it is integrable or by numerical integration, and eqn (14.7) is fitted to the observed pattern by adjusting parameters in $f(x)_K$ just as in the case of individual profile fitting. In this approach, several functions have been used for $f(x)$ and $g(x)$. They are a single Lorentzian and the sum of seven Lorentzians (Taupin 1973; Parrish et al. 1976), Pearson VII and observed profile (Toraya 1988), and the pseudo-Voigt and the pseudo-Voigt convoluted by an exponential function (Enzo et al. 1988), respectively.

The following is an example of the second approach (Toraya 1988). By replacing the integration in eqn (14.7) with the numerical quadrature,

$$y(x_i) = b(x_i) + \sum_K \sum_j f(x_i - x_j)_K g(x_j)_K \, \Delta x. \tag{14.8}$$

The Pearson VII function $P_{P7}(x)$ is used to express $f(x)_K$ in the form, $f(x)_K \cong I_K P_{P7}(x)_K$. Then the derivative of eqn (14.8) with respect to the lth adjustable parameter p_1 in $f(x)_K$ can be obtained by

$$\partial y(x_i)/\partial p_1 = \sum_j \partial f(x_i - x_j)_K/\partial p_1 g(x_j)_K \, \Delta x. \tag{14.9}$$

Thus the function (14.8) can be least-squares fitted to the observed diffraction profile by using the Gauss–Newton method. Figure 14.2 shows (a) the observed diffraction profile of ZnO (102) used as $g(x)$, (b) a fitting result using the function (14.8), and (c) $f(x)_K$ profiles for two overlapping reflections (112/200) of 4 mol% Y_2O_3-doped tetragonal ZrO_2. In this profile fitting, the two reflections were assumed to have the same profile shape (one FWHM parameter and one exponent parameter in Pearson VII), while the positional and intensity parameters of the two reflections were refined independently (Table 14.3).

The two approaches mentioned above can be used for deriving the intrinsic diffraction profiles $f(x)$ from overlapping reflections. The methods carry out, at the same time, pattern decomposition. It may also be important to notice that the method using the fitting function (14.8), in which the a priori determined and fixed instrumental profile $g(x)$ is convoluted with an adjustable profile used for the intrinsic diffraction profile $f(x)$, has a higher resolving power for overlapping profiles than does the commonly-used

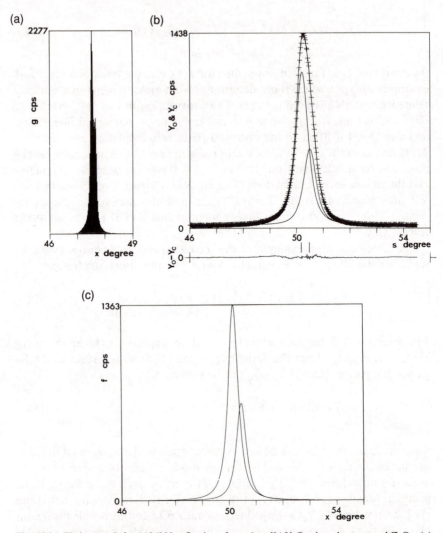

Fig. 14.2 Fitting result for 112/200 reflections from 4 mol% Y_2O_3-doped tetragonal ZrO_2: (a) $g(x)$ profile before normalization (102 reflection from ZnO), (b) observed and calculated profile intensities and their difference (the diagram is shown as in Fig. 14.1). Individual $h(x)_K$ profiles are also shown, and (c) resolved $f(x)$ profiles.

profile fitting based on the function (14.4), in which both instrumental and specimen profiles are modelled together by a single profile function. One reason may be that the modelling of complex diffraction features such as asymmetry is absorbed into the pre-determined $g(x)$ function, while the function $f(x)$ with a simple form (usually symmetric) gives a sharp minimum to the residual function in the least-squares fitting. This approach, extended

Table 14.3 Refined least-squares parameters, R_p and R_{wp} factors for 112 and 200 reflections from 4 mol% Y_2O_3-doped tetragonal ZrO_2

b_0		21.9(23)	
b_1		0.0005(5)	
hkl	112		200
I_K	812(12)		402(13)
t_K (°)	50.182(3)		50.475(5)
H_K (°)		0.436(5)	
m		1.31(1)	
R_p (%)		1.9	
R_{wp} (%)		3.5	

to the whole powder pattern, is discussed again in Section 14.4.3. Chapter 7 is largely based on this approach and reports excellent results for modelling asymmetry and for determining crystallite size and microstrain parameters.

14.4 Applications of the Pawley method

14.4.1 *Refinement of unit-cell parameters*

The procedure for precise determination of unit-cell parameters with the powder method is well established for materials of high crystallographic symmetry. One would not expect such high precision from the powder diffraction patterns of high peak density and/or poorly resolved peaks because of indexing ambiguities. Therefore, refinement of unit-cell parameters of materials with complex powder diffraction patterns is an important application of the Pawley method.

Most computer programs for Pawley and Rietveld methods can refine the parameter for a 2θ-zero correction together with those for the unit-cell. However, the peak maximum is shifted to the low-angle side due to the profile asymmetry, which is particularly pronounced in the case of conventional X-ray diffractometers, and the amount of the peak shift is generally angle-dependent. The procedure here presented for correcting systematic peak shift is based on the internal standard technique. It uses the computer program WPPF, which can decompose the powder pattern of a multi-component mixture (Toraya 1986). The WPPF incorporates two species functions. One is a generalized polynomial function for peak-shift correction, which is given by

$$\Delta 2\theta = \sum_m t_m \Phi_m^{m-1} \qquad (14.10)$$

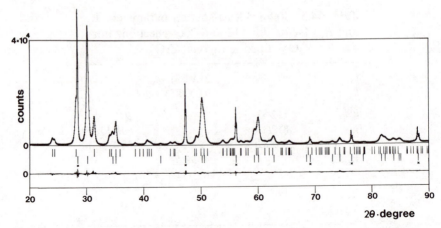

Fig. 14.3   Whole-powder-pattern fitting result for a mixture of 2.5 mol% Y_2O_3-doped mono-clinic and tetragonal ZrO_2 and Si standard. The diagram is shown as in Fig. 14.1.

where the t_m are the adjustable parameters in the least-squares refinement and Φ_m may be either 2θ or $\tan\theta$. The peak position is corrected by $x_K^{corrected} = x_K + \Delta 2\theta$. During the least-squares fitting, the unit-cell parameters of the sample to be investigated and parameters t_m in (14.10) are varied, while the unit-cell parameters of the standard reference material are fixed. Then the peak shift is automatically corrected against the unit-cell parameters of standard reference material. The other device is the scale factor, by which the integrated intensities I_K's of 'each phase' in a multi-component mixture are multiplied.

The usefulness of scale factor is shown in the following example, which is the refinement of unit-cell parameters of Y_2O_3-doped monoclinic and tetragonal ZrO_2 powders (Toraya 1989a). The small crystallite size and microstrain of ZrO_2 gave poor angular resolution to the powder diffraction patterns (Fig. 14.3). With increasing Y_2O_3 content (from 0 to 7.8 mol% Y_2O_3), the amount of tetragonal phase increased, while that of monoclinic phase decreased and finally became a trace amount. The Pawley refinement was started with the sample of the pure monoclinic phase, and could be forwarded to the samples with successively less monoclinic content. A set of refined integrated intensity parameters of the previous sample was used as the starting parameters of the least-squares refinement for the next sample. When the relative amount of the monoclinic phase was low and the integrated intensities of the monoclinic phase were too weak to be varied independently, then the scale factor for the monoclinic phase was adjusted instead of the individual integrated intensity parameters. Figure 14.3 shows a whole pattern fitting result from application of the Pawley method to the powder data of a three-phase mixture (2.5 mol% Y_2O_3-doped monoclinic

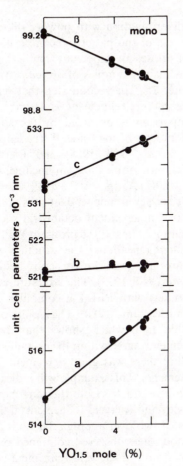

Fig. 14.4 Variation of unit cell parameters with $YO_{1.5}$ for monoclinic ZrO_2.

and tetragonal ZrO_2, and the Si standard). Figure 14.4 shows the plot of refined unit-cell parameters of monoclinic ZrO_2 as a function of composition. Even with poorly resolved data as shown in Fig. 14.3, the plot has a smaller scatter of individual values, and thus gives a more reliable result, than did a previous study which used the conventional technique of measuring peak positions (Ruh *et al.* 1984).

14.4.2 *Ab initio structure determination and structure refinement*

Use of the Pawley method produces an output list of indexed integrated intensities. These data can be used for the *ab initio* structure determination followed by structure refinement based on the integrated intensity method.

Success in structure determination with powder diffraction data is largely dependent on the success of the pattern decomposition, and therefore the Pawley refinement constitutes an important step in the structure analysis. A general procedure and many examples of *ab initio* structure determination are reviewed in Chapter 15. The second step that may be made with the integrated intensities, structure refinement, is described below.

An interesting comparison is provided by the structure refinement of $Na_2Al_2Ti_6O_{16}$, the isomorph of the mineral freudenbergite, by both 'two-stage' analysis (Will *et al.* 1983, 1988) and Rietveld refinement. The compound has the crystallographic data: monoclinic, $C2/m$, $a = 12.1239(3)$, $b = 3.7749(1)$, $c = 6.4180(2)$ Å, $\beta = 107.59(1)°$, $V = 280.00(4)$Å^3, $Z = 1$ (Toraya *et al.* 1990). It has seven independent sites in the unit cell, and Al^{3+} and Ti^{4+} ions share two independent octahedral sites. Pawley refinement with synchrotron radiation data and the program WPPF produced intensity values for 94 independent reflections. The structure was then refined with POWLS, a least-squares program for structure refinement with powder integrated-intensity data (Will 1979). The structure was also refined (from the same powder data set) with PFLS, a computer program for Rietveld refinement (Toraya and Marumo 1980). The results of the two refinements are compared in Table 14.4, which shows that the atomic parameters obtained in both refinements agree within the estimated standard deviations. It is also important that there was good agreement in the site occupancy factor in the two refinements. This example will indicate that the maximum information which can be extracted from the same powder data set, can be the same for both Pawley and Rietveld refinements if there are no problems. It should be noted that if two overlapping reflections could not be resolved with the Pawley method, their observed combined profile would have the same weight as one reflection in Rietveld refinement also.

The Rietveld method is more straightforward than the 'two-stage' analysis, if the purpose of the analysis is the structure refinement. However, if the calculation of Fourier series is required to see the electron density distribution, the Pawley method becomes important again. Most computer programs for the Rietveld method can also provide the integrated intensities of individual reflections in the last cycle of the least-squares, which are calculated from the formula provided by Rietveld (Rietveld 1969). However, we must be careful in using these data since they are biased by the structure model refined by the Rietveld method. Recently, the maximum entropy method has been shown to be a powerful technique for obtaining the electron density distribution from the powder diffraction data (Sakata *et al.* 1990). In that method, the accurate measurement of integrated intensities is also necessary if one is to obtain an unbiased electron density map. It seems safely predictable that the Pawley method will become more important in electron-density distribution studies.

Table 14.4 Atomic parameters of $Na_2Al_2Ti_6O_{16}$ refined under the same conditions with both two-stage analysis and Rietveld refinement techniques

		Two-stage analysis	Rietveld refinement
Ti(1)	g	0.67(1)	0.670(8)
	x	0.2989(4)	0.2996(2)
	z	0.7081(9)	0.7082(4)
	B	0.43(22)	0.53(9)
Al(1)	g	0.33	0.33
Ti(2)	g	0.87(1)	0.874(8)
	x	0.3973(4)	0.3975(2)
	z	0.2959(3)	0.2954(3)
	B	0.80(19)	0.78(7)
Al(2)	g	0.13	0.126
Na	g	0.92	0.908
	B	1.77(59)	1.47(21)
O(1)	x	0.3690(9)	0.3678(5)
	z	0.9939(20)	0.9933(9)
	B	$-0.18(22)$	1.00(14)
O(2)	x	0.2347(9)	0.2352(4)
	z	0.3451(17)	0.3451(7)
	B	-0.18	1.07(16)
O(3)	x	0.1361(10)	0.1361(4)
	z	0.7065(18)	0.7068(7)
	B	-0.18	0.02(15)
O(4)	x	0.4435(11)	0.4443(5)
	z	0.6378(15)	0.6377(7)
	B	-0.18	1.41(16)
R_B (%)		5.0	6.2

14.4.3 *Lattice-direction-dependent crystallite size and microstrain*

The parameters concerning crystallite size and microstrain can be refined along with the unit-cell (and structural) parameters by using whole-powder-pattern fitting. This idea was pointed out at an early stage of the development of the Rietveld method (Young *et al.* 1977), and has been advanced in Rietveld refinement. The procedure based on the use of the Voigt function (Langford 1978) was first outlined by De Keijser *et al.* (1983), and some works appeared along this line (Cox 1984; Hill and Howard 1985; David and Matthewman 1985; Thompson *et al.* 1987). Young and Desai (1988) reviewed these studies, and, starting from the work by Thompson *et al.* (1987), further formulated the expression for FWHM parameters in the pseudo-Voigt function. In their procedure, the FWHM parameters related to the instrumental broadening were pre-determined by using the profiles of standard reference material, and the remaining FWHM parameters, which are relevant to crystallite size and microstrain, were least-squares fitted together with those for structural parameters. The reader will notice the common feature between this procedure and the first approach mentioned in Section 14.3.2: the Voigt function has been extended to microstructural analysis by the Rietveld method.

A procedure which corresponds to the second approach in Section 14.3.2, has also been developed (Howard and Snyder 1984; Le Bail 1984). This procedure, using numerical convolution, relies on the power of a computer rather than the useful fact of the Voigt function. It has, however, another merit. For example, the profile modelling is independent of various factors in the particular experimental conditions, and the work can be concentrated on modelling the $f(x)$ profile. In both approaches, some works incorporated the anisotropy of crystal imperfection into the profile modelling (Greaves 1985; Thompson et al. 1987; Lartigue et al. 1987; Lutterotti and Scardi 1990). The analysis of crystallite size and microstrain, however, does not normally require calculation of structure factors and, therefore, the task is well suited to the Pawley method.

The following example is an application of the Pawley method to crystallite size and microstrain analysis, where the lattice-direction dependent profile width is incorporated into the profile model (Toraya 1989b). The procedure is based on the fitting function by eqn (14.8) and the integral breadth method. Either a pseudo-Voigt function or a Pearson VII function was used to represent the intrinsic diffraction profile $f(x)$. Experimental diffraction profiles of a specimen which shows no appreciable broadening were chosen at various 2θ positions in order to provide the angle variation of profile shape, and they were used to construct the instrumental function $g(x)$. The FWHM parameter H, in the case of the pseudo-Voigt function, is expressed with the integral breadth β and the mixing factor, η, as

$$H_{pv} = [\eta 2/\pi + (1 - \eta)2(\ln 2/\pi)^{\frac{1}{2}}]\beta. \tag{14.11}$$

In one model, isotropy was assumed for crystallite size and microstrain effects as well as Lorentzian size and microstrain profiles, and β was expressed by (Wilson 1962)

$$\beta = 4/3 \, \lambda/D_s \cos \theta + 4\varepsilon \tan \theta, \tag{14.12}$$

where D_s is a sphere diameter and ε is a microstrain parameter. The other model, discussed in this section, incorporates anisotropic size effects with the assumption of cylindrical crystallite shape, of which the two limiting cases are acicular and disk-like forms. The equations for calculating integral breadth Scherrer constants for cylindrical shapes, derived by Langford and Louër (1982), are

$$\beta = \lambda/(h \cos \theta) \qquad \text{for } \Psi = 0,$$

$$\beta = (\pi \sin \Psi/D_c)[\tfrac{8}{3} + 2q \cos^{-1} q - (\sin^{-1} q/2q)$$
$$\quad - \tfrac{5}{2}(1 - q^2)^{1/2} + \tfrac{1}{3}(1 - q^2)^{3/2}]^{-1}(\lambda/\cos \theta) \qquad \text{for } 0 < \Psi \leq \alpha,$$

$$\beta = (\pi \sin \Psi/D_c)[\tfrac{8}{3} - (\pi D_c \cot \Psi)/4h]^{-1}(\lambda/\cos \theta) \qquad \text{for } \alpha < \Psi \leq \pi/2,$$

$$\alpha = \tan^{-1}(D_c/h) \qquad \text{and} \qquad q = h/D_c \tan \Psi \qquad (14.13)$$

where D_c and h are the cylinder diameter and height, respectively, and Ψ is the acute angle between the cylindrical axis and scattering vector. The FWHM parameter H is a function of 2θ angle by the combination of eqns (14.11) and (14.12), while it is mainly a function of hkl when eqn (14.13) is substituted into eqn (14.11). D_s and ε are adjustable parameters in the case of eqn (14.12), as are D_c and h in eqn (14.13). The fitting function eqn (14.8) is least-squares fitted in a manner similar to the procedure described in Section 14.3.2 but with more sophisticated computation for saving computer memory and computation time (Toraya 1989b).

The following is a test result for this procedure in which the sample was hydrothermally formed hexagonal hydroxyapatite, $Ca_5(PO_4)_3OH$ ($a = 9.424$ and $c = 6.881$ Å). The observed individual particle shape was columnar as seen via transmission electron microscopy (TEM). The observed diffraction profiles of CeO_2 (NIST standard reference material 674) were used as the standard to provide $g(x)$. The powder data ($2\theta = 20-80°$) were least-squares fitted under the assumptions of two crystallite models, one cylindrical and the other spherical. The corresponding two refinement results, C and D, are compared in Table 14.5 and Fig. 14.5. Other refined parameters, not presented in Table 14.5, were background parameters, integrated intensity parameters for all reflections, unit-cell parameters, 2θ-zero, and the

Table 14.5 Refined least-squares parameters, R_p and R_{wp} factors for hydroxyapatite, $Ca_5(PO_4)_3OH$

Refinement	Crystallite model	D_c (Å)	h (Å)	h/D_c	η	R_p (%)	R_{wp} (%)
C	Cylinder	327(2)	943(16)	2.9	0.586(7)	4.2	6.3
D	Sphere		461(6)	1.0	0.68(2)	8.8	12.9

(a)

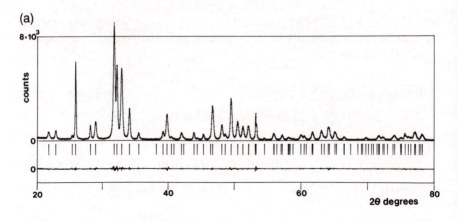

(b)

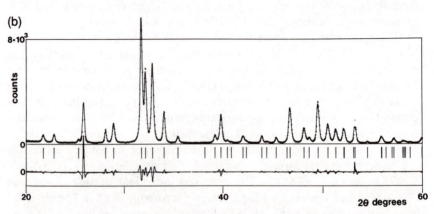

Fig. 14.5 (a) Fitting result for hydroxyapatite, $Ca_5(PO_4)_3OH$, obtained by assuming the cylindrical crystallite crystallite model. (b) A part of the fitting result for the same sample as in (a) but refined with the spherical crystallite model. The diagram is shown as in Fig. 14.1

parameter η in the pseudo-Voigt function. The refinement C, in which columnar crystallites were assumed, gives an excellent fit between the observed and calculated intensities, as is indicated by the R_p and R_{wp} factors, and the deduced crystallite sizes, D_c and h, agreed with TEM observation. On the other hand, the spherical crystallite model gave large misfits for the reflections with scattering vectors in the directions both along the column axis and perpendicular to it.

14.5   Conclusion

The individual profile-fitting, Pawley, and Rietveld methods have been applied to many studies in crystallography and the field of materials science. They have been successful in providing new information about the structure of materials and have also contributed to increasing the power and range of powder-data analyses. For the study of materials, these three methods are better used complementarily to each other rather than on their own. For example, individual profile-fitting is important in the initial stage of data analysis, when we have little knowledge about the material under investigation. The Pawley method is applied after the unit-cell parameters are known approximately. It can also be used to check extinction rules for deriving possible space groups and detecting extra peaks from impurities. The derived integrated intensities are used for *ab initio* structure determination, calculation of Fourier maps, and structure refinement, along with the use of the Rietveld method.

No structural model is required for the individual profile-fitting and Pawley methods in calculating the profile intensity. This is sometimes a great advantage because we can, for example, refine the unit-cell parameters of the materials with a high peak density in their powder patterns without knowledge of their crystal structures (Section 14.4.1). Microstructural analysis such as of crystallite size and microstrain may also be better done via the Pawley method rather than the Rietveld method (14.4.3), since the former technique does not include errors arising from the structural model and preferred orientation correction. The lack of a requirement for a structural model also reduces the time and cost to obtain results.

The Pawley method uses whole pattern data in its analysis, as does the Rietveld method, and a high precision is expected from it in deriving unit-cell and microstructural parameters. The profile shape, which is designed to be angle-dependent and also lattice-direction-dependent in the case of microstructural analysis for anisotropic crystal imperfection, is extrapolated to the angular regions of low resolution and weak reflections. This position-constrained pattern decomposition method exhibits its efficiency in decomposing complex powder patterns and deriving useful parameters from them. Examples have been shown in Sections 14.4.1–14.4.3. The whole-powder-pattern fitting analysis is rapid and straightforward compared to the 'two-stage' analysis via individual profile fittings, and can be used almost automatically for samples analysed routinely.

14.6 Acknowledgement

The author is grateful to Dr W. Parrish for helpful discussions in preparing the manuscript.

References

Brown, A. and Edmonds, J. W. (1980). *Adv. X-ray Anal.*, **23**, 361–74.
Caglioti, G., Paoletti, A., and Ricci, F. P. (1958). *Nucl. Instrum.*, **3**, 223–8.
Cox, D. E. (1984). *Acta Crystallogr.*, **A40S**, C369.

David, W. I. F. and Matthewman, J. C. (1985). *J. Appl. Crystallogr.*, **18**, 461–6.

De Keijser, Th. H., Langford, J. I., Mittemeijer, E. J., and Vogels, A. B. P. (1982). *J. Appl. Crystallogr.*, **15**, 308–14.

De Keijser, Th. H., Mittemeijer, E. J., and Rozendaal, H. C. F. (1983). *J. Appl. Crystallogr.*, **16**, 309–16.

Enzo, S., Fagherazzi, G., Benedetti, A., and Polizzi, S. (1988). *J. Appl. Crystallogr.*, **21**, 536–42.

Greaves, C. (1985). *J. Appl. Crystallogr.*, **18**, 48–50.

Hall Jr., M. M., Veeraraghavan, V. G., Rubin, H., and Winchell, P. G. (1977). *J. Appl. Crystallogr.*, **10**, 66–8.

Hepp, A. and Baerlocher, Ch. (1988). *Austral. J. Phys.*, **41**, 229–36.

Hill, R. J. and Howard, C. J. (1985). *J. Appl. Crystallogr.*, **18**, 173–80.

Howard, C. J. (1982). *J. Appl. Crystallogr.*, **15**, 615–20.

Howard, S. A. and Snyder, R. L. (1984). *Acta Crystallogr.*, **A40S**, C369.

Huang, T. C. and Parrish, W. (1975). *Appl. Phys. Lett.*, **27**, 123–4.

Jansen, E., Schäfer, W., and Will, G. (1988). *J. Appl. Crystallogr.*, **21**, 228–39.

Khattak, C. P. and Cox, D. E. (1977). *J. Appl. Crystallogr.*, **10**, 405–11.

Langford, J. I. (1978). *J. Appl. Crystallogr.*, **11**, 10–14.

Langford, J. I. (1987). *Prog. Cryst. Growth Charact.*, **14**, 185–211.

Langford, J. I. and Louër, D. (1982). *J. Appl. Crystallogr.*, **15**, 20–6.

Langford, J. I., Louër, D., Sonneveld, E. J., and Visser, J. W. (1986). *Powder Diffract.*, **1**, 211–21.

Lartigue, C., Le Bail, A., and Percheron-Guegan, A. (1987). *J. Less Com. Met.*, **129**, 65–76.

Le Bail, A. (1984). *Acta Crystallogr.*, **A40S**, C369.

Louër, D., Auffrédic, J. P., Langford, J. I., Ciosmak, D., and Niepce, J. C. (1983). *J. Appl. Crystallogr.*, **16**, 183–91.

Louër, D., Weigel, D., and Louboutin, R. (1969). *Acta Crystallogr.*, **A25**, 335–8.

Lutterotti, L. and Scardi, P. (1990). *J. Appl. Crystallogr.*, **23**, 246–52.

Mackie, P. E. and Young, R. A. (1975). *Acta Crystallogr.*, **A31**, S198.

Malmros, G. & Thomas, J. O. (1977). *J. Appl. Crystallogr.*, **10**, 7–11.

Mortier, W. J. and Constenoble, M. L. (1973). *J. Appl. Crystallogr.*, **6**, 488–90.

Moraweck, B., De Montgolfier, Ph., and Renouprez, A. J. (1977). *J. Appl. Crystallogr.*, **10**, 184–90.

Parrish, W., Huang, T. C., and Ayers, G. L. (1976). *Trans. Am. Crystallogr. Assoc.*, **12**, 55–73.

Pawley, G. S. (1981). *J. Appl. Crystallogr.*, **14**, 357–61.

Rietveld, H. M. (1969). *J. Appl. Crystallogr.*, **2**, 65–71.

Ruh, R., Mazdiyasni, K. S., Valentine, P. G., and Bielstein, H. O. (1984). *J. Am. Ceram. Soc.*, **67**, C190–2.

Sakata, M., Mori, R., Kumazawa, S., Takata, M., and Toraya, H. (1990). *J. Appl. Crystallogr.*, **23**, 526–34.

Scott, H. G. (1987). Abstract of *Int. Symp. X-ray Powder Diffractometry*, Fremantle, 88.

Sonneveld, E. J. and Visser, J. W. (1975). *J. Appl. Crystallogr.*, **8**, 1–7.

Stokes, A. R. (1948). *Proc. Phys. Soc. Lond.*, **61**, 382–91.

Taupin, D. (1973). *J. Appl. Crystallogr.*, **6**, 266–73.

Thompson, P., Cox, D. E., and Hastings, J. B. (1987). *J. Appl. Crystallogr.*, **20**, 79–83.

Thompson, P., Reilly, J. J., and Hastings, J. M. (1987). *J. Less Com. Met.*, **129**, 105–14.

Toraya, H. (1986). *J. Appl. Crystallogr.*, **19**, 440–7.

Toraya, H. (1988). *J. Appl. Crystallogr.*, **21**, 192–6.

Toraya, H. (1989*a*). *J. Am. Ceram. Soc.*, **72**, 662–4.

Toraya, H. (1989*b*). *Powder Diffract.*, **4**, 130–6.

Toraya, H. (1990). *J. Appl. Crystallogr.*, **23**, 485–91.

Toraya, H. and Marumo, F. (1980). *Rep. Res. Lab. Engng. Mater. Tokyo Inst. Technol.*, **5**, 55–64.

Toraya, H. and Parrish, W. (1990). *Acta Crystallogr.*, **A46S**, C-62.

Toraya, H., Masciocchi, N., and Parrish, W. (1990). *J. Mater. Res.*, **5**, 1538–43.

Voigt, W. (1912). *Sitzungsber. K. Bayer. Akad. Wiss.*, **42**, 603–20.

Warren, B. E. (1969). *X-ray diffraction.* Addison-Wesley, Reading, MA.

Wertheim, G. K., Butler, M. A., West, K. W., and Buchanan, D. N. E. (1974). *Rev. Sci. Instrum.*, **45**, 1369–71.

Will, G. (1979). *J. Appl. Crystallogr.*, **12**, 483–5.

Will, G., Parrish, W., and Huang, T. C. (1983). *J. Appl. Crystallogr.*, **16**, 611–22.

Will, G., Masciocchi, N., Parrish, W., and Hart, M. (1987). *J. Appl. Crystallogr.*, **20**, 394–401.

Will, G., Bellotto, M., Parrish, W., and Hart, M. (1988). *J. Appl. Crystallogr.*, **21**, 182–91.

Wilson, A. J. C. (1962). *X-ray optics.* Methuen, London.

Wilson, A. J. C. (1963). *Mathematical theory of X-ray powder diffractometry.* Gordon and Breach, London.

Windsor, C. G. and Sinclair, R. N. (1976). *Acta Crystallogr.*, **A32**, 395–409.

Young, R. A. and Desai, P. (1988). *Arch. Nauk. Mater.*, **10**, 71–90.

Young, R. A. and Wiles, D. B. (1982). *J. Appl. Crystallogr.*, **15**, 430–8.

Young, R. A., Mackie, P. E., and von Dreele, R. B. (1977). *J. Appl. Crystallogr.*, **10**, 262–9.

Young, R. A., Prince, E., and Sparks, R. A. (1982). *J. Appl. Crystallogr.*, **15**, 357–9.

15

Ab initio structure solution with powder diffraction data

Anthony K. Cheetham

15.1 Introduction

There is a long and well-established tradition of using powder diffraction methods to study crystal structures. The earliest examples were inevitably concerned with the structures of simple materials such as iron metal (Hull 1917), and although for many decades powder methods were used predominantly as a means of qualitative analysis, a number of structure determinations, e.g. α- and β-UF$_5$ (Zachariasen 1949), were reported. Such studies were largely based upon geometrical considerations and trial-and-error methods, although Zachariasen and Ellinger (1963) were able to solve the monoclinic structure of β-plutonium by a manual direct-methods procedure. Nevertheless, powder methods were largely restricted to studies on simple structure types until the advent of the Rietveld method in 1969 (Rietveld 1969) extended their scope to the *refinement* of complex, low symmetry systems with as many as 50 atoms in the asymmetric unit (Cheetham and Taylor 1977; Hewat 1986). This important development has inevitably led to a re-examination of the methods that are available for the *solution* of unknown structures, and it is with this exciting challenge that the present chapter is concerned. We stress, however, that the Rietveld method continues to play a vital role in the subsequent refinement, which is necessary in order to complete the structure determination.

A number of indirect methods have been successfully applied to the determination of structures from powder data, as described in Section 15.4, below, but recent advances in both instrumentation and computational techniques have brought us to the point where it is now becoming possible to tackle this problem in a systematic manner (Christensen *et al.* 1985; Rudolf and Clearfield 1985; Cheetham 1986a). Such approaches are known as *ab initio* techniques. On the instrumental side, improvements in resolution, both with synchrotron X-ray and time-of-flight neutron methods, have led to dramatic reductions in peak overlap (which is a major source of ambiguity in obtaining integrated intensity information). At the same time, advances in computational techniques, for example the use of negative quartets in direct methods, are also making a significant contribution. In the present work, we review the achievements to date and examine some of the areas in which future progress may be anticipated. These include the use of anomalous scattering with synchrotron X-rays, the utilization of more sophisticated direct-methods procedures, and the application of computer modelling techniques. A number of examples, based upon both constant wavelength and time-of-flight neutron diffraction, and both laboratory and synchrotron X-ray diffraction, are described.

Powder diffraction methods play a central role in the discovery and characterization of new materials. Ideally, a new phase is characterized structurally by single crystal X-ray diffraction, but for materials that can only be prepared in the polycrystalline form, or cannot readily be studied under the required conditions of, say, temperature or pressure, the crystallographer is forced to rely on powder techniques. We should recall, however, that this difficult task is rarely approached empty-handed. A wide range of advanced physical methods is now available for the study of powders, and many of these can provide useful preliminary information (Cheetham 1986b). Knowledge of the chemical composition of the material is obviously an asset, and can usually be obtained by classical methods or by analytical electron microscopy (Cheetham and Skarmulis 1981). Selected area electron diffraction might yield information about the crystallographic unit cell (Wright *et al.* 1985a), and magic-angle spinning NMR may, in certain cases (e.g. carbon, phosphorus, aluminium, silicon, and platinum), reveal the number of atoms of a particular element in the asymmetric unit (Thomas *et al.* 1983). Other useful methods include IR and Raman spectroscopies, which provide information relating to the symmetry of particular groups within a structure, and Mössbauer spectroscopy, which may yield data on the site symmetry and oxidation state of, say, iron atoms. Such complementary data should preferably be obtained prior to embarking on a structure determination by powder diffraction methods.

15.2 Structure determination—an overview

The determination of a crystal structure, whether by single crystal or powder methods, can usefully be described in terms of a series of discrete steps:

1. Indexing of the diffraction pattern and the determination of the crystal system (e.g. cubic, orthorhombic, etc.) and lattice parameters.

2. Identification of the space group.

2. Solution of the phase problem and determination of an approximate structure.

4. Refinement of the structure.

With single crystal X-ray data that have been collected on an automatic diffractometer, this sequence of steps is normally routine. The unit cell can be identified from an automatic peak search, the space group determined from systematic absences, and the phase problem solved with Patterson or direct methods. Patterson methods are normally used if there is one or more dominant scatterers present, as in many organometallic and coordination compounds, and direct methods are particularly suitable for organic compounds, minerals, and other continuous solids. The structure can then be refined by using a combination of least-squares and difference Fourier techniques. With powders, however, there are serious difficulties, most of which arise because the three-dimensional intensity data is compressed into a single dimension in the powder experiment. This results in peak overlap, which introduces complications at each stage of the structure determination.

15.3 Unit cell determination

The first hurdle to be overcome, viz. indexing of the reflections and unit cell determination, is non-trivial, although our recent experience has been that few determinations now falter at this stage. Several excellent auto-indexing programs are available, including ITO (Visser 1969), TREOR (Werner *et al.* 1985), and DICVOL (Louër and Vargas 1982). They require very accurate *d*-spacing data, but these can normally be obtained by peak-fitting of individual reflections. Careful measurement of the zero-point error in the counter setting is essential and can most conveniently be carried out by means of a preliminary scan with a well-characterized standard. Clearly the auto-indexing procedure is greatly facilitated if high resolution data are available, as in the following examples which were determined (Wilkinson and Cheetham, unpublished results) from synchrotron powder data of the quality illustrated in Fig. 15.1. The estimated standard deviations obtained from least-squares refinements are given in parentheses:

$Cu_6Mo_5O_{18}$: $a = 15.2556(7)$, $b = 6.2748(4)$, $c = 14.701(1)$ Å, $\beta = 101.87(1)°$.   Volume $= 1377.19$ Å^3.

Cimetidine: $a = 10.213(1)$, $b = 17.966(2)$, $c = 6.5147(9)$ Å, $\beta = 111.28(1)°$.   Volume $= 1113.85$ Å^3.

$Zn_3[Fe(CN)_6]_2$: $a = 12.604(1)$, $c = 32.950(3)$; rhombohedral. Volume $= 4533.05$ Å^3.

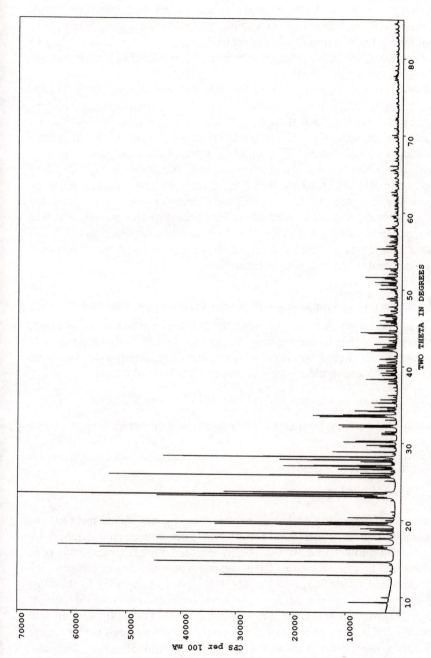

Fig. 15.1 A high resolution, synchrotron X-ray pattern of the drug cimetidine, collected at Daresbury Laboratory.

Selected-area electron diffraction is of value in obtaining the unit cell in difficult cases, since the high magnification of the transmission electron microscope permits the single crystallites that comprise a powder sample to be interrogated individually. This approach can also be useful in helping to identify the space group, though care must be exercised in taking account of double diffraction effects.

Following the determination of the unit cell, the solution of the crystal structure may still present formidable obstacles. In particular, the solution of the phase problem is likely to be difficult, largely because the overlapping of adjacent reflections introduces ambiguity into the assignment of intensities to particular *hkl* values. This problem arises both with low symmetry structures, where the density of reflections may become very high, and with high symmetry structures, in which there may be unavoidable overlap of non-equivalent reflections. Unambiguous intensity data are needed for both Patterson and direct-methods calculations. For this reason, structure determination using powders has, until very recently, remained an empirical, rather than an exact, science. Some of the approaches that have been used are described in the following section.

15.4 Structure determination with powder data: empirical methods

It is important to bear in mind that the Rietveld method is a refinement procedure and that its successful implementation requires that a reasonable starting model should be available. A number of strategies have been used to obtain starting models, once the unit cell has been obtained:

1. Identification of an isostructural material of known structure.

2. Use of difference Fourier methods to investigate derivatives of known structures.

3. Trial and error methods.

4. Computer modelling techniques.

Prior to attempting to solve the structure by any other method, the possibility that an isostructural material has already been characterized, i.e. method (1), above, should be thoroughly explored. Structures that have been determined in this way range from systems where the analogy is obvious, for example $Cr_2(MoO_4)_3$ and $Fe_2(MoO_4)_3$ (Battle *et al.* 1985), to those for which a certain amount of detective work is necessary, as in the case of Ba_2ReO_5, which adopts the same structure as Cs_2DyCl_5 (Cheetham and Thomas 1987). A powerful way of carrying out such a search is to examine an appropriate database, e.g. the Inorganic Crystal Structure Database (ICSD), for a compound with similar stoichiometry and lattice parameters to the unknown. In our experience, this procedure can frequently eliminate a great deal of unnecessary work.

The difference Fourier approach, method (2), has also been extensively exploited, but it suffers from the fact that the magnitudes of individual observed I_{hkl} values are in many cases uncertain due to overlapping of reflections. However, when most of the atomic positions have been located and refined, the Rietveld procedure provides a list of approximate observed I_{hkl}'s which can be used with the calculated I_{hkl}'s in a difference Fourier analysis. This approach should be used with caution, but it has been applied successfully to some interesting problems in which a parent structure is already well-known. These include the location of adsorbed molecules in zeolite cages (Fitch *et al.* 1985; Wright *et al.* 1985*b*), and the positions of protons in complex oxides (Rotella *et al.* 1982). Strategies based upon derivatives of known structures are enormously valuable in solid state chemistry, but like method (1), they are obviously unsuitable for the characterization of entirely new structure types.

For new structures, workers have hitherto been forced to rely heavily upon trial-and-error approaches (method (3)). These have often involved the use of both X-ray and neutron data, the former being used to determine heavy atom positions from radial distribution functions, and the latter to locate lighter atoms and to provide a precise refinement of the structure by Rietveld methods. The work of Taylor on uranium tetrabromide provides a good example; uranium was located with X-rays and the lighter halogen atoms were positioned by trial-and-error using neutron data (Taylor and Wilson 1974). In other instances, this methodology was used to solve the non-centrosymmetric superstructure of Bi_3ReO_8 (Cheetham and Rae-Smith 1985) and an ordered structure in the cerium hydride system (Titcomb *et al.* 1974). An interesting example of the trial-and-error approach is found in the case of zeolite ZSM-23 (Wright *et al.* 1985*a*), in which the lattice parameters were obtained by electron diffraction and the basic building units of the structure were deduced from adsorption data. The units were then manipulated manually in order to fit into the unit cell.

The trial-and-error approach to structure solving, whilst sometimes unavoidable, is extremely time-consuming and frequently unrewarding, but attempts are now being made to automate the procedure in order to eliminate the possibility that viable models might be overlooked in a manual search. Unlike human beings, computers do not become bored when looking through a large number of possibilities! This automated trial-and-error procedure we have labelled as computer modelling (method (4)). An illustration can be found in a recent study of lanthanum palladium oxides (Attfield 1988), in which the metal positions were initially found by means of a novel, computer search procedure based upon interatomic distances. Oxygen atoms were then placed by inspection. Another example is to be found in recent work on zeolites in which a Monte Carlo sampling method was used to predict the possible arrangements of *n* tetrahedral Si/Al (T) sites within a unit cell of known dimensions (Deem and Newsam 1989). The value of *n* may be obtained from adsorption data or ^{29}Si magic-angle spinning NMR results, and the space group possibilities should be apparent from the diffraction data. Within the symmetry constraints, the Si/Al atoms can then be arranged according to the likely distance between adjacent T sites and the probable T—T—T angle. Models that are compatible with the above constraints can then be optimized by simulated annealing and their powder patterns calculated with, say, LAZYPULVERIX. In a variation on this theme, we are aware from a personal communication that de Bruijn and van Mechelen at Shell, Amsterdam, are using computer graphics with LAZY-PULVERIX to solve molecular structures with known unit cells. Modelling techniques of this general type should be capable of revealing the structures of a wide range of materials in which the stereochemistry of the basic structural elements is well understood.

In spite of the undoubted power of the above methods, however, there remains a significant body of structural problems that do not lend themselves to any of the empirical approaches. In the following section we describe some of the new developments that enable structures to be solved from powder data in a systematic manner with no prior knowledge of the structure.

15.5   *Ab initio* structure determination with powder data

15.5.1   *Methodology*

The *ab initio* determination of crystal structures from powder data involves a series of steps that mimics the sequence of events in a single crystal study:

1. Determination of the unit cell.

2. Decomposition of powder pattern into a series of integrated intensities.

3. Assignment of space group from systematic absences.

4. Solution of the phase problem by Patterson or direct methods.

5. Refinement of structure by Rietveld analysis.

The unit cell determination is normally performed by auto-indexing of the diffraction pattern, as described in Section 15.3, above. The next step is to use the unit cell in order to extract the maximum amount of integrated intensity data (I_{hkl}'s) from the powder diffraction pattern. This pattern decomposition is discussed in detail in Chapter 14. It is perhaps the most difficult stage in the determination, because peak overlap will lead to uncertainties in the I_{hkl} values of adjacent ot fully overlapping Bragg reflections. Pawley (1981) has developed a very effective way of performing this step, by means of a least-squares analysis in which the unit cell dimensions, zero-point error, peak shape parameters and I_{hkl} values are varied (Section 14.4). It should be stressed that no structural model is required for this procedure; it is not a Rietveld refinement. Ambiguities arising from partial or complete overlap of reflections are taken into account in the e.d.s.'s of the individual I_{hkl} values. An example of such an analysis is shown in Fig. 15.2. It should be clear that this crucial step can be performed more effectively with high resolution data, a theme to which we shall return in Section 15.5.3.

From this stage onwards, the analysis mirrors that of a single crystal study. The possible space groups can be assigned from the systematic absences, although in cases of uncertainty it may be useful to carry out Pawley refinements in a number of alternative space groups, or to obtain a series of electron diffraction patterns. The phase problem is solved by conventional crystallographic methods. The principal difficulties are that the data set will be considerably smaller than that obtained from a single crystal and that some of the data will be unreliable for the reasons discussed above. It is a tribute to the robustness of modern structure-solving techniques that it is still possible to determine structures under these circumstances. In the following sections, we examine a number of examples that have been carried out with laboratory X-rays, synchrotron X-rays, and neutron data.

15.5.2   Structure determination with laboratory X-rays

The first attempt to solve a structure from powder data by a systematic *ab initio* approach appears to be the study of β-plutonium by Zachariasen and Ellinger (1963). The pattern decomposition step was achieved by collecting data at a series of temperatures and relying on the anisotropy of the cell expansion to separate partially overlapping Bragg reflections. This ingenious strategy could still prove useful, even with modern, high resolution diffractometer data. The phase problem was then solved by a *manual* direct-methods analysis to reveal seven independent plutonium atoms in the asymmetric unit of the monoclinic cell ($a = 9.284$, $b = 10.643$, $c = 7.859$ Å, $\beta = 92.13°$, space group $I2/m$).

Several groups have subsequently addressed the problem of solving crystal structures from laboratory X-ray powder data. For inorganic materials, the use of Patterson techniques rather than direct methods is often preferred, and again the main problem lies in the acquisition of a sufficiently large set of unique I_{hkl} values. Berg and Werner (1977) illustrated the practicability of this approach with a study of the coordination compound $(NH_4)_4[(MoO_2)_4O_3]\cdot(C_4H_3O_5)_2$ in which X-ray data were collected on a Guinier–Hägg focusing camera using Cu K_α radiation; peak positions and

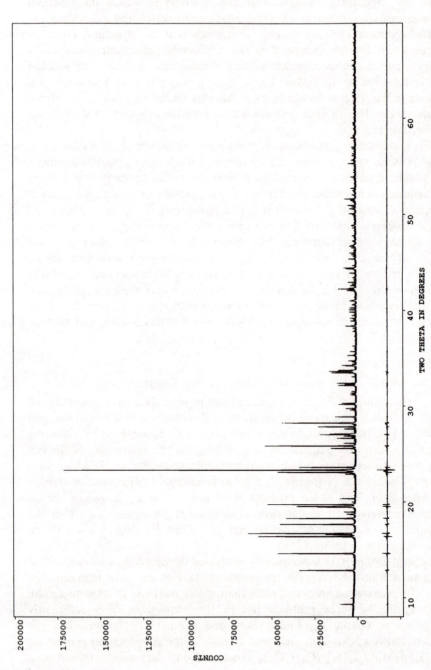

Fig. 15.2 The results of a Pawley analysis of the data shown in Fig. 15.1. The observed profile is shown as dashes and the calculated profile as a smooth curve; a difference profile is also presented. The analysis yielded over 600 allowed reflections, of which only 134 gave E-values > 1.2.

Table 15.1 Some examples of *ab initio* structure determinations from powder data

Compound	Space group	No. of atoms in asymmetric unit	Ref.
(a) Laboratory data			
β-Pu	$I2/m$	7	Zachariasen and Ellinger (1963)
$(NH_4)_4[(MoO_2)_4O_3] \cdot (C_4H_3O_5)_2$	$C2$	16	Berg and Werner (1977)
$CaCl_2(NH_3)_n$	$Abm2(n = 2)$; $Pnma(n = 8)$	$5(n = 2)$; $7(n = 8)$	Westerman et al. (1981)
$ZrKH(PO_4)_2$	$P2/c$	15	Rudolf and Clearfield (1984)
$Nd(OH)_2NO_3 \cdot H_2O$	$C2/m$	9	Louër and Louër (1987)
$KCaPO_4 \cdot H_2O$	$C2/m$	7	Louër et al. (1988)
$Cd_3(OH)_5(NO_3)$	$Pmmn$	9	Plévert et al. (1989)
α-$VO(HPO_4) \cdot 2H_2O$	$P2_1/c$	14*	Le Bail et al. (1989a)
$NaPbFe_2F_9$	$C2/c$	9	Le Bail (1989b)
β-$VO(HPO_4) \cdot 2H_2O$	$P1$	18	Le Bail et al. (1989)
(b) Synchrotron X-rays			
α-$CrPO_4$	$Imma$	8	Attfield et al. (1986)
I_2O_4	$P2_1/c$	6	Lehmann et al. (1987)
$MnPO_4 \cdot H_2O$	$C2/c$	6	Lightfoot et al. (1987)
PbC_2O_4	$P1$	7	Christensen et al. (1989)
Clathrasil, sigma-2	$I4_1/amd$	17	McCusker (1988)
$LaMo_5O_8$	$P2_1/a$	14	Hibble et al. (1988)
$LiCH_3$	$I222$	11	Weiss et al. (1990)
(c) Neutrons			
$FeAsO_4$	$P2_1/n$	6	Cheetham et al. (1986)

* Including 5 hydrogen atoms located by neutron diffraction.

intensities were measured with a microdensitometer. In a contrasting approach, Rudolf and Clearfield (1984) collected data on a conventional X-ray powder diffractometer to solve the crystal structure of $ZrKH(PO_4)_2$. A number of similar studies have recently appeared in the literature, as shown in Table 15.1(a).

It is clear, however, that the full potential of the X-ray method is more likely to be realized at synchrotron sources, where it is possible to achieve an extraordinary level of angular resolution in the diffraction pattern. This approach is addressed in the following section.

15.5.3 *Structure determination with synchrotron X-rays*

Synchrotron X-rays have several profound advantages over conventional X-rays for structure determination. First, the combination of high brightness and excellent vertical collimation can be harnessed to construct diffractometers with unparalleled resolution, as in the case of Cox's instrument at the National Synchrotron Light Source (NSLS), Brookhaven National Laboratory (Cox *et al.* 1986), where the resolution at the focusing position is $<0.02°$ in 2θ. Second, the *tunability* of synchrotron radiation opens up new opportunities to utilize anomalous scattering in structure determination with powders. With high resolution the problem of peak overlap is reduced to a minimum, which permits the maximum amount of unambiguous I_{hkl} data to be obtained from the diffraction pattern. Pawley and Rietveld analysis can be more difficult at high resolution because, in many instances, sample effects such as strain or particle size may dominate the peak shape. Nevertheless, Cox *et al.* (1983) have shown that these refinement procedures can be applied to such data and several successful structure determinations have recently been performed (Table 15.1(b)). In the first example, by Attfield *et al.* (1986), the orthorhombic structure of $\alpha\text{-CrPO}_4$, with eight atoms in the asymmetric unit, was solved by Patterson methods using a vector search procedure; 68 well-resolved peaks were utilized. A relatively poor R_{wp}-factor (19.3 per cent) was obtained for the final Rietveld refinement with the synchrotron data, no doubt due to problems with preferred orientation and *hkl*-dependent line broadening effects, but a subsequent medium resolution neutron study (on D1a at ILL Grenoble) gave an excellent fit ($R_{wp} = 8.3$ per cent), confirming the correctness of the X-ray model. A comparison (Table 15.2) of the coordinates obtained from the X-ray and neutron refinements, and a subsequent single crystal study (Glaum *et al.* 1986), nicely illustrates the relative strengths of the different methods (Attfield *et al.* 1988).

A somewhat more complex example (Hibble *et al.* 1988) will serve to illustrate the inherent difficulties that are encountered when structure solution is attempted with powder data. High resolution data from a sample of a new cluster compound, LaMo_5O_8, were collected on the Cox instrument at the NSLS. The unit cell dimensions ($a = 9.914$, $b = 9.089$, $c = 7.573$ Å, $\beta = 109.08°$) were found by auto-indexing, and the space group ($P2_1/c$) was obtained from the systematic absences, albeit not without some difficulty due to the presence of impurity lines. The data-set contained 297 Bragg reflections, of which it was possible, by means of a Pawley analysis, to assign unambiguous I_{hkl} values in 191 cases. Of these, however, 148 had zero intensity, leaving only 43 I_{hkl} values with significant intensity! Nevertheless, it proved possible to locate the heavy atoms (one La and five Mo in the asymmetric unit) by direct methods with the aid of negative quartets, which harness the information content of the weak intensity data (Schenk 1982).

Table 15.2 Structural parameters for α-CrPO$_4$ refined using synchrotron X-ray (marked X) and neutron (N) data in *Imma* (No. 74) with e.d.s.'s in parentheses. Values from the reported single crystal study[a] (marked S) are given for comparison

Atom	x	y	z	B_{iso}[b]
Cr(1)	1/4	1/2	0	0.3(2)N
				0.283(6)S
Cr(2)	1/4	0.3660(3)X	1/4	
		0.3650(4)N		0.0(1)N
		0.36611(3)S		0.316(4)S
P(1)	1/2	1/4	0.0819(12)X	
			0.0790(8)N	0.0(1)N
			0.0825(2)S	0.30(1)S
P(2)	1/4	0.5738(4)X	1/4	
		0.5739(2)N		0.47(8)N
		0.57358(5)S		0.345(7)S
O(1)	0.3790(10)X	1/4	0.2269(17)X	
	0.3766(3)N		0.2280(5)N	0.53(8)N
	0.3773(2)S		0.2268(3)S	0.42(2)S
O(2)	0.3603(6)X	0.4914(5)X	0.2145(11)X	
	0.3610(2)N	0.4907(1)N	0.2142(3)N	0.62(6)N
	0.3611(1)S	0.4902(1)S	0.2146(2)S	0.42(1)S
O(3)	0.2263(6)X	0.6352(5)X	0.0576(10)X	
	0.2240(1)N	0.6368(2)N	0.0546(3)N	0.68(5)N
	0.2238(1)S	0.6363(1)S	0.0552(2)S	0.56(1)S
O(4)	1/2	0.3509(8)X	−0.0457(15)X	
		0.3486(2)N	−0.0422(4)N	0.31(7)N
		0.3496(2)S	−0.0432(3)S	0.50(2)S

[a] Glaum *et al.* (1986).
[b] For the powder X-ray refinement, overall $B_{iso} = 0.24(7)$ Å^2.

The oxygen atoms, eight in all, were then located by difference Fourier methods. As in several other examples (e.g. Table 15.2), the precision of the final X-ray Rietveld refinement is rather modest. One of the reasons for this is clearly that the high-angle intensities are diminished by the X-ray form factor, unlike the corresponding neutron patterns, but in addition there is reason to believe that the quality of the fit often falls short of expectations. This can probably be ascribed to the comparative difficulty of collecting an X-ray data set that is free of systematic errors, in particular preferred orientation or graininess. Further effort is required in this direction in order to improve the quality of high-resolution X-ray refinements.

15.5.4 *Structure determination with neutrons*

The possibility of collecting ultra-high-resolution powder data is not confined to the synchrotron X-ray technique. The new generation of time-of-flight powder diffractometers, particularly at pulsed neutron sources, are also very impressive, especially when they are operated in the back-scattering mode with a long flight path. For example, a resolution of $\Delta d/d = 0.04$ per cent can be obtained on the High Resolution Powder Diffractometer, HRPD, at the ISIS Spallation Neutron Source in the UK. The feasibility of solving structures from neutron powder data has been discussed by Christensen *et al.* (1985). Direct-methods, rather than Patterson analysis, are better suited to neutron data because neutron scattering lengths fall within a rather narrow range of values. Patterson methods find favour in systems containing a small number of strong scatterers which dominate the vector map, a situation that is more typically encountered with X-rays.

One of the early experiments on HRPD (Cheetham *et al.* 1986) demonstrated what might be done with the new generation of neutron instruments. The material studied was $FeAsO_4$, which was first reported by Shafer *et al.* (1956). Subsequently, D'Yvoire indexed the X-ray powder pattern according to a monoclinic cell and assigned the space group $P2_1/n$, and on the basis of IR evidence and the facile transformation of the monoclinic modification to one with the $CuSO_4$ structure, it was suggested that the iron atom might be octahedrally coordinated (D'Yvoire 1972). Data for $FeAsO_4$ were collected on HRPD at ISIS in the high resolution mode and the correctness of the cell proposed by D'Yvoire was confirmed with auto-indexing. The lattice parameters were then refined and the space group confirmed by examining the systematic absences. In the absence, at the time, of a Pawley-type programme for time-of-flight data, integrated intensities were obtained manually for 139 reflections, including approximately 60 weak reflections. Structure factor amplitudes were then used as the input for a direct-methods analysis, in which, as in the case of $LaMo_5O_8$, both triplets and negative quartets were calculated.

The direct methods peak list from the calculation with the highest figure of merit is shown in Table 15.3. Analysis of the interpeak distances and angles confirmed that a chemically sensible solution had been found, with peaks 1 and 2 corresponding to Fe and As, respectively, and peaks 3–6 to O. Note that the peaks appear in order of their neutron scattering lengths and that there is a substantial gap between peaks 6 and 7; the latter indicates the level of noise in the map. The coordinates obtained from the direct methods analysis were then refined by integrated intensity methods to yield a final

Table 15.3 (a) Peak listing for $FeAsO_4$ from direct methods solution[a]

Peak No.	X	Y	Z	Height
1	0.178	0.449	0.758	2691
2	0.075	0.179	0.243	1960
3	0.210	0.080	0.413	1230
4	0.015	0.373	0.369	1199
5	0.898	0.046	0.178	1193
6	0.131	0.256	0.913	1104
7	0.562	0.587	0.726	591
8	0.385	0.617	0.887	590

(b) Final atomic coordinates for $FeAsO_4$, with e.s.d.'s in parentheses[b]

Atom	X	Y	Z
Fe	0.173(2)	0.462(2)	0.763(2)
As	0.073(3)	0.202(2)	0.223(4)
O(1)	0.257(3)	0.101(3)	0.420(4)
O(2)	0.027(3)	0.377(2)	0.384(4)
O(3)	0.089(4)	0.077(3)	0.152(4)
O(4)	0.125(4)	0.267(3)	0.939(5)

[a] No. of reflections = 139. Triplets generated from top 66 reflections. Negative quartets generated from top 67 reflections. Figure of merit = 4.0.
[b] Overall temperature factor = $-3.8(4)$ Å^2.

R-factor, $R(I) = 6.5$ per cent. One of the interesting features of the structure of $FeAsO_4$ is the presence of five coordinated iron sites which share a common edge.

It is perhaps significant that this early work on HRPD has not been succeeded by a number of similar examples. In the majority of cases it will be easier to solve a structure from X-ray rather than neutron data, simply because with X-rays the phase problem can normally be solved on the basis of locating a sub-set of the atoms in the asymmetric unit (i.e. the heavier atoms). With neutrons it will usually be necessary to locate a majority of the atoms before the structure factors can be reliably phased, thus limiting the complexity of structure that can be tackled with a reasonable prospect of success. It seems likely, therefore, that the role of neutrons in powder studies will focus more on their use to obtain precise atomic coordinates, a feature that is underlined by the comparison shown in Table 15.2.

15.6 Future prospects

The results that have been obtained by a variety of methods have demonstrated that structure determination from powder data is now a viable option in circumstances in which single crystal measurements are not possible. All of the structures solved to data have been relatively simple, and there remains a substantial gap between the complexity of the structures that can be solved and those that can be refined by the Rietveld method. Taking a direct-methods solution, for example, a simple rule of thumb is that 10 E-values of reasonable magnitude are required per atom in the asymmetric unit. For the solution of a 30 atom structure, therefore, we would require substantially in excess of 300 unique I_{hkl} values, bearing in mind that many of them are likely to yield E's that are too small to contribute significantly to the E map. On the face of it this target seems rather daunting, but the use of negative quartets as well as triplets offers some encouragement because it harnesses the information content of the smaller E-values (Schenk 1982). This strategy was certainly crucial in the work on $FeAsO_4$.

Other approaches that are being explored include the use of maximum entropy methods for the structure solution stage (Henderson and Gilmore 1989), and a more imaginative utilization of Patterson techniques (Altomare *et al.* 1992); the latter, for example, might be harnessed for the pattern decomposition stage, given that a properly decomposed profile ought to yield a Patterson map that is positive throughout. Another opportunity is presented by the ever-increasing power of modern computers, which enables one to undertake lengthy calculations that would not have been feasible in the past. In this laboratory, for example, we have recently located all 17 'heavy' atoms in the (known) structure of the drug cimetidine, $C_{10}N_6SH_{16}$, using an automatic sequence of least-squares and $2F(\text{obs}) - F(\text{calc})$ maps, and taking as the starting point a rather unpromising E map based upon only 134 E values (> 1.2) (Cernik *et al.* 1991).

Ultra-high-resolution powder data are clearly destined to play a major role in structure determination, but early experiences of analysing such data have shown that the non-structural characteristics of the sample now play a pivotal role in the success or otherwise of the experiment. The peak shape is the major problem, because subtleties in the line shape, due to microstrain, stacking faults, and anisotropy in the crystallite size, are no longer obscured by the instrumental contribution to the peak width. The crystallite size is especially important in time-of-flight work because it influences both the linewidth and the degree of extinction (which is wavelength dependent); the ideal size appears to be about 1 μm (Cernik *et al.* 1991). The variation of line broadening, not only with d-spacing but also from one class of *hkl* reflections to another, requires careful study and is continuing to hinder the optimal use of pattern decomposition routines and Rietveld refinement in structure determination. Nevertheless, the prospect of solving structures of substantial complexity is well within reach and will present an exciting challenge during the coming decade.

References

Altomare, A., Cascarano, G., and Giacovazzo, C. (1992). *Acta Crystallogr.*, **A48**, 30.

Attfield, J. P. (1988). *Acta Crystallogr.*, **B44**, 563.

Attfield, J. P., Sleight, A. W., and Cheetham, A. K. (1986). *Nature*, **322**, 620.

Attfield, J. P., Cheetham, A. K., Cox, D. E., and Sleight, A. W. (1988). *J. Appl. Crystallogr.*, **21**, 452.

Battle, P. D., Cheetham, A. K., Harrison, W. T. A., Pollard, N. J., and Faber, J. (1985). *J. Solid State Chem.*, **58**, 221.

Berg, J.-E. and Werner, P.-E. (1977). *Z. Kristallogr.*, **145**, 310.

Cernik, R. J., Cheetham, A. K., Prout, C. K., Watkin, D. J., Wilkinson, A. P., and Willis, B. T. M. (1991). *J. Appl. Crystallogr.*, **24**, 222.

Cheetham, A. K. (1986a). *Mater. Sci. Forum*, **9**, 103.

Cheetham, A. K. (1986b). *Proc. Indian Nat. Sci. Acad.*, **52A**, 25.

Cheetham, A. K. and Rae-Smith, A. R. (1985). *Acta Crystallogr.*, **B41**, 225.

Cheetham, A. K. and Skarnulis, A. J. (1981). *Anal. Chem.*, **53**, 1060.

Cheetham, A. K. and Taylor, J. C. (1977). *J. Solid State Chem.*, **21**, 253.

Cheetham, A. K. and Thomas, D. M. (1987). *J. Solid State Chem.*, **71**, 61.

Cheetham, A. K., David, W. I. F., Eddy, M. M., Jakeman, R. J. B., Johnson, M. W., and Torardi, C. C. (1986). *Nature*, **320**, 46.

Christensen, A. N., Lehmann, M. S., and Nielsen, M. (1985). *Austral. J. Phys.*, **38**, 497.

Christensen, A. N., Cox, D. E., and Lehmann, M. S. (1989). *Acta Chem. Scand.*, **43**, 19.

Cox, D. E., Hastings, J. B., Thomlinson, W., and Prewitt, C. (1983). *Nucl. Instrum. Methods*, **208**, 573.

Cox, D. E., Hastings, J. B., Cardoso, L. P., and Finger, L. W. (1986). *Mater. Sci. Forum*, **9**, 1.

Deem, M. W. and Newsam, J. M. (1989). *Nature*, **342**, 260.

D'Yvoire, F. C. (1972). *Compt. Rendu Hebd. Séanc. Acad. Sci., Paris*, **275C**, 949.

Fitch, A. N., Jobic, H., and Renouprez, A. (1985). *J. Chem. Soc., Chem. Commun.*, 284.

Glaum, R., Gruehn, R., and Moller, M. (1986). *Z. Anorg. Allg. Chem.*, **543**, 111.

Henderson, K. and Gilmore, C. J. (1989). In *Maximum entropy and Bayesian methods* (ed. J. Skilling), p. 233.

Hewat, A. W. (1986). *Chem. Scripta*, **26A**, 119.

Hibble, S. J., Cheetham, A. K., Bogle, A. R. L., Wakerley, H. R., and Cox, D. E. (1988). *J. Am. Chem. Soc.*, **110**, 3295.

Hull, A. W. (1917). *Phys. Rev.*, **9**, 84.

Le Bail, A. (1989). *J. Solid State Chem.*, **83**, 267.

Le Bail, A., Ferey, G., Amoros, P., and Beltran-Porter, D. (1989a). *Eur. J. Solid State Inorg. Chem.*, **26**, 419.

Le Bail, A., Ferey, G., Amoros, P., Beltran-Porter, D., and Villeneuve, G. (1989b). *J. Solid State Chem.*, **79**, 169.

Lehmann, M. S., Christensen, A. N., Fjellvag, H., Feidenhans'l, R., and Nielsen, M. (1987). *J. Appl. Crystallogr.*, **20**, 123.

Lightfoot, P., Cheetham, A. K., and Sleight, A. W. (1987). *Inorg. Chem.*, **26**, 3544.

Louër, D. and Louër, M. (1987). *J. Solid State Chem.*, **68**, 292–9.

Louër, D. and Vargas, R. (1982). *J. Appl. Crystallogr.*, **15**, 542.

Louër, M., Plévert, J., and Louër, D. (1988).*Acta Crystallogr.*, **B44**, 463–7.

McClusker, L. B. (1988). *J. Appl. Crystallogr.*, **21**, 305.

Pawley, G. S. (1981). *J. Appl. Crystallogr.*, **14**, 357.

Plévert, J., Louër, M., and Louër, D. (1989). *J. Appl. Crystallogr.*, **22**, 470–5.

Rietveld, H. M. (1969). *J. Appl. Crystallogr.*, **2**, 65.

Rotella, F. J., Jorgensen, J. D., Biefeld, R. M., and Morosin, B. (1982). *Acta Crystallogr.*, **B38**, 1697.

Rudolf, P. and Clearfield, A. (1984). *Inorg. Chem.*, **23**, 4679.

Rudolf, P. and Clearfield, A. (1985). *Acta Crystallogr.*, **B41**, 418.

Sabine, T. M. (1985). *Austral. J. Phys.*, **38**, 507.

Schenk, H. (1982). In *Computational crystallography* (ed. D. Sayre), p. 65. Oxford University Press.

Shafer, E. C., Shafer, M. W., and Roy, R. (1956). *Z. Kristallogr.*, **108**, 263.

Taylor, J. C. and Wilson, P. W. (1974). *Acta Crystallogr.*, **B30**, 2664.

Thomas, J. M., Klinowski, J., Ramdas, S., Hunter, B. K., and Tennakoon, D. T. B. (1983). *Chem. Phys. Lett.*, **102**, 158.

Titcomb, C. G., Cheetham, A. K., and Fender, B. E. F. (1974). *J. Phys. C: Solid State Phys.*, **7**, 2409.

Visser, J. W. (1969). *J. Appl. Crystallogr.*, **2**, 89.

Weiss, E., Corbelin, S., Cockcroft, J. K., and Fitch, A. N. (1990). *Angew. Chem. Int. Edn. Engl.*, **29**, 650.

Werner, P.-E., Eriksson, L. and Westdahl, M. J. (1985). *J. Appl. Crystallogr.*, **18**, 367.

Westerman, S., Werner, P.-E., Schuler, T., and Raldow, W. (1981). *Acta Chem. Scand.*, **A35**, 467.

Wilkinson, A. P. and Cheetham, A. K. Unpublished results.

Wright, P. A., Thomas, J. M., Millward, G. R., Ramdas, S., and Barri, S. A. I. (1985a). *J. Chem. Soc., Chem. Commun.*, 1117.

Wright, P. A., Thomas, J. M., Cheetham, A. K., and Nowak, A. K. (1985b). *Nature*, **318**, 611.

Zachariasen, W. H. (1949). *Acta Crystallogr.*, **2**, 296.

Zachariasen, W. H. and Ellinger, F. H. (1963). *Acta Crystallogr.*, **16**, 369.

Index